AF800264

Offshore Wind Farm Technology

Yongqian Liu · Shuang Han · Jie Yan

Offshore Wind Farm Technology

中国工信出版集团

電子工業出版社
PUBLISHING HOUSE OF ELECTRONICS INDUSTRY
http://www.phei.com.cn

Springer

Yongqian Liu
School of New Energy
North China Electric Power University
Beijing, China

Shuang Han
School of New Energy
North China Electric Power University
Beijing, China

Jie Yan
School of New Energy
North China Electric Power University
Beijing, China

ISBN 978-981-96-1888-0 ISBN 978-981-96-1889-7 (eBook)
https://doi.org/10.1007/978-981-96-1889-7

Jointly published with Publishing House of Electronics Industry
The print edition is not for sale in China (Mainland). Customers from China (Mainland) please order the print book from: Publishing House of Electronics Industry.

© Publishing House of Electronics Industry 2025

This work is subject to copyright. All rights are solely and exclusively licensed by the Publisher, whether the whole or part of the material is concerned, specifically the rights of reprinting, reuse of illustrations, recitation, broadcasting, reproduction on microfilms or in any other physical way, and transmission or information storage and retrieval, electronic adaptation, computer software, or by similar or dissimilar methodology now known or hereafter developed.
The use of general descriptive names, registered names, trademarks, service marks, etc. in this publication does not imply, even in the absence of a specific statement, that such names are exempt from the relevant protective laws and regulations and therefore free for general use.
The publishers, the authors, and the editors are safe to assume that the advice and information in this book are believed to be true and accurate at the date of publication. Neither the publishers nor the authors or the editors give a warranty, express or implied, with respect to the material contained herein or for any errors or omissions that may have been made. The publishers remain neutral with regard to jurisdictional claims in published maps and institutional affiliations.

This Springer imprint is published by the registered company Springer Nature Singapore Pte Ltd.
The registered company address is: 152 Beach Road, #21-01/04 Gateway East, Singapore 189721, Singapore

If disposing of this product, please recycle the paper.

Contents

Chapter 1
Introduction

1.1 Outline

Vigorously developing renewable energy has become a key measure to deal with world energy problems such as climate change, fossil energy resource depletion, and environmental pollution. It is the only way for the sustainable development of human society and nature. The world is rich in wind energy resources. In the past 10 years, wind power technology has made remarkable progress, and the cost of wind power has dropped rapidly. It has become the primary power source of power systems in many countries.

The industrial scale and technological progress of China's wind power are at the world's first-class level. By the end of 2020, the cumulative installed capacity of wind power in China reached 280 GW, accounting for 12.8% of the installed capacity in the power system. Onshore wind power technology is becoming more mature, and the cost of power generation has achieved parity with coal power. Wind power will play a significant role in the country's goal of addressing climate change by achieving carbon peaking in 2030 and carbon neutrality in 2060. It will become essential to China's future new power system with new energy as the main body.

At present, the total installed capacity of China's onshore wind power is close to 300 GW, and high-quality resources are dwindling. Accelerating the research and development of offshore wind power technology and industrial development has become one of the important directions of China's energy transformation.

1.2 Key Features of Offshore Wind Power

Compared with onshore wind power, offshore wind power has apparent advantages:

© The Author(s), under exclusive license to Springer Nature Singapore Pte Ltd. 2025

Y. Liu et al., *Offshore Wind Farm Technology*,
https://doi.org/10.1007/978-981-96-1889-7_1

(1) Do not occupy scarce onshore land resources. With the continuous expansion of the scale of wind farms, land resources for onshore wind power development are becoming increasingly scarce. Offshore wind power does not occupy land and benefits from vast, continuous sea space, which is ideal for large-scale wind power projects.
(2) Close to the load center, which is convenient for power consumption. The phenomenon of wind and electricity curtailment is a major problem in onshore wind power in China. The main reason is that the system's peak-shaving capacity is seriously insufficient, and the reverse distribution of resources and loads has become a rigid constraint restricting new energy consumption. Offshore wind power is close to the coastal power load center, which can realize nearby and local consumption and reduce transmission costs.
(3) Little environmental impact on residents. When the wind turbine runs, it will produce noise, electromagnetic wave pollution, etc. It does not matter to worry about the noise and electromagnetic waves generated by offshore wind turbines and the impact of wind turbines on residents' vision. Offshore wind turbines can even run at high speeds to improve power generation efficiency.
(4) Offshore wind energy resources are abundant. The offshore wind speed is high, the calm period is short, and the wind power utilization hours are high. Research data shows that compared with onshore wind farms, the offshore wind speed 10 km offshore is usually 25% higher than that on the coast; the average annual utilization hours of onshore wind power is about 2200 h, and the average annual utilization hours of offshore wind power can reach 3000 h above.
(5) Offshore wind shear is relatively small. The roughness of the sea is much smaller than that of the land, resulting in small wind shear, which can obtain the same wind speed at a lower height. It can reduce the height of the wind turbine tower and save the cost of construction and installation.
(6) Wind fluctuations are smaller at sea. The temperature difference between day and night and the temperature between seasons at sea is much smaller than that at land, so the turbulent intensity of sea air is relatively small. The low turbulence intensity can reduce the stress variability of the wind turbines, which is more beneficial to the fatigue life of the blades, towers, and basic structures of the wind turbine.

Of course, in addition to the above advantages, there are still many difficulties and challenges in developing offshore wind power. Offshore wind power is very different from onshore wind power in terms of design, installation, operation, and maintenance, and requires expensive special equipment and installation ships. Fundamental investment, operation, and maintenance costs for offshore wind turbines remain high. Offshore wind power is affected by the natural environment (waves, salt erosion, rough wind, seabed, etc.), bringing new technical challenges. Offshore wind power technologies, including planning, design, construction, operation and maintenance, etc., still need to be mature and are currently the critical direction of international wind energy science and technology research.

1.3 Current Situation of Offshore Wind Power Development

The inception of global offshore wind power can be traced back to Europe, with Denmark leading the charge through the establishment of the Vindeby offshore wind farm in 1991. This pioneering project, comprising 11 wind turbines each with a capacity of 450 kW, has been operational for three decades. In recent years, driven by concerns over climate change and the shift towards low-carbon energy sources, the offshore wind power industry has gained significant momentum worldwide. Currently, over 50 countries and regions have embarked on offshore wind power development.

According to data from the Global Wind Energy Council (GWEC), global offshore wind power has experienced remarkable growth, averaging a 24% annual increase since 2013. This surge has resulted in a cumulative installed capacity that has leaped from 4 GW in 2011 to 35 GW in 2020, as illustrated in Fig. 1.1.

The primary growth in offshore wind power has been observed in the European market, with the UK and Germany as key players, and the Asian market, notably led by China, as depicted in Table 1.1 and Fig. 1.2, respectively. While Europe added 2.9 GW of offshore wind power capacity in 2020, marking a 20% decrease from 2019 due to factors such as COVID-19, China has sustained a robust growth trajectory. Conversely, the American market (primarily North America) has exhibited slower progress. Nevertheless, amidst the global acceleration of offshore wind power development, the Americas, as an emerging market, hold promise for substantial growth in the foreseeable future.

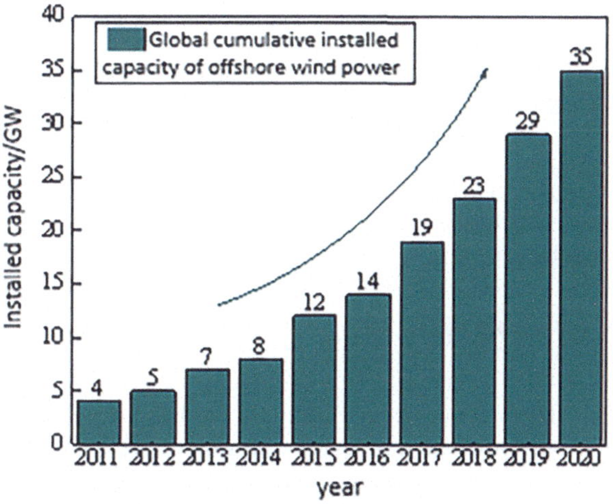

Fig. 1.1 Cumulative installed capacity of global offshore wind power from 2011 to 2020 (*Source* GWEC)

Table 1.1 Statistical table of new installed capacity of offshore wind power in different countries or regions from 2016 to 2020 (*Source* GWEC, Wind Europe)

Countries		Year				
		Newly installed capacity of offshore wind power/MW				
		2016	2017	2018	2019	2020
European	The UK	56	1715	1312	1764	483
	Germany	849	1253	969	1111	219
	Other European countries	659	228	380	752	2216
Asian	China	592	1161	1655	2395	3060
	Other Asian countries	37	115	35	123	60
America	The US	30	0	0	0	12

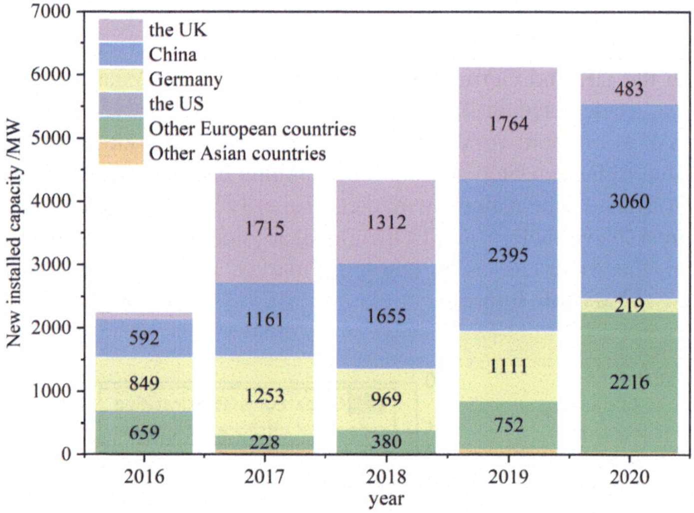

Fig. 1.2 Newly installed capacity of offshore wind power in different countries or regions from 2016 to 2020. *Note* The data label only labels the data that accounts for more than 3% of the overall data bar

1.3.1 Offshore Wind Power in Europe

The offshore wind power industry has been developed in Europe for nearly 30 years. It is currently one of the most important forms of renewable energy power generation in Europe, and it also plays a vital role in planning future renewable energy. In the early 1990s, Denmark, the Netherlands, the United Kingdom, and other European countries were the first to carry out offshore demonstration projects to verify the feasibility of offshore wind power technology. During this period, offshore wind power's overall construction scale and stand-alone capacity were relatively small.

According to statistics, from 1991 to 2001, European countries built 9 offshore wind power projects in total. Among them, 5 projects have a capacity of less than 10 MW, and the total investment is no more than 100 million EUR.

In 2002, under the background that the governments of various countries gradually recognized the development prospects of offshore wind power and provided various policy support; European offshore wind power entered a rising period of commercial development. During the period, the average construction scale of a single project reached 400 MW, and the cumulative installed capacity exceeded 6 GW. Offshore wind turbines entered the era of high power, with the average unit power reaching 4 MW, and the average cost of electricity dropped to 0.69 ~ 1.29 yuan/kW h. The investment scale exceeded 2 billion EUR.

Through these projects, leading developers have accumulated rich experience in design, construction, operation, and maintenance. The cost of electricity for offshore wind power in Europe has gradually decreased. Since 2012, the government has shifted from buying electricity at high prices to bidding for power plants, and Europe has officially opened the era of offshore wind power parity. At present, the electricity cost of offshore wind power in Europe has been lower than 0.5 yuan/kW h. The bidding price of offshore wind power in the UK has dropped to 0.35 yuan/kW h. Germany has also achieved zero subsidies. Most of the European projects planned to be implemented in 2023–2025 have electricity prices below 0.4 yuan/kW h.

According to the latest statistics released by Wind Europe, in 2020, the new offshore installed capacity in Europe will be 2.9 GW: 1493 MW in the Netherlands, 706 MW in Belgium, 483 MW in the United Kingdom, 219 MW in Germany, and 17 MW in Portugal. By the end of 2020, the total installed capacity of offshore wind power in Europe was 25 GW, including 5402 grid-connected wind turbines in 12 countries. Among them, the UK has the largest installed offshore wind power capacity in Europe (10.2 GW) and is the largest installed offshore wind power capacity in the world, accounting for about 42% of all installed capacity (see Fig. 1.3). Germany comes in second with 31%, followed by the Netherlands (10%), Belgium (9%), and Denmark (7%).

1.3.2 Offshore Wind Power in China

Compared with Europe, China's offshore wind power started relatively late. In November 2007, China's first offshore wind power test project—CNOOC's offshore drilling and platform test unit (1.5 MW) was completed and put into operation, marking a "zero breakthrough" in China's offshore wind power development. In 2010, China's first and Asia's first large plastic sea Shanghai Donghai Bridge offshore wind farm demonstration project (100 MW) was connected to the grid for power generation, marking that China has mastered the construction technology of offshore wind power projects. With the gradual maturity of equipment technology and continuous accumulation of development experience, China's offshore wind power has gradually entered a period of accelerated development under the vigorous promotion

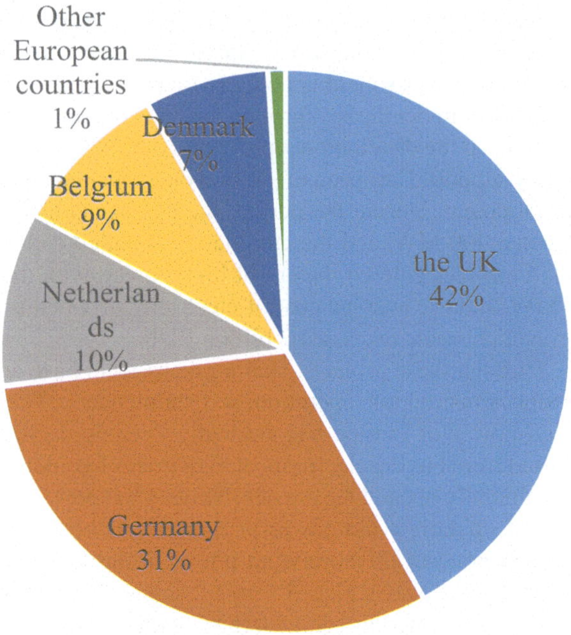

Fig. 1.3 Market share of cumulative installed capacity of offshore wind power in European countries (*Source* Wind Europe)

of relevant policies at the national and local government levels: 2014 was called the first year of offshore wind power by the industry and experienced explosive growth. The new installed capacity reached 230 MW, a year-on-year increase of 283%. In 2015 and 2016, it entered a stage of rapid development, with newly added installed capacities of 360 MW and 590 MW and growth rates of 57% and 64%, respectively.

In November 2016, the National Energy Administration officially issued the "Thirteenth Five-Year Plan for Wind Power Development", which proposed to ensure that 5 GW of offshore wind power will be connected to the grid by 2020. The cumulative installed capacity of wind power will reach more than 210 GW, focusing on promoting Offshore wind power construction in Jiangsu, Zhejiang, Fujian, Guangdong, and other provinces. As a result, China's offshore wind power has officially entered the fast lane of development, and the bidding for installed capacity has been fully launched. The stages and key policies of China's offshore wind power are shown in Table 1.2.

According to the latest data released by GWEC, the newly installed capacity of global offshore wind power in 2020 was 6.1 GW. Among them, China's newly installed capacity is about 3 GW, accounting for over half of the world's newly installed capacity. This is the third consecutive year that China has ranked first in the world in terms of the newly installed capacity of offshore wind power. By the end of 2020, China's cumulative installed capacity of offshore wind power has reached 9.9 GW, making it the second largest offshore wind power market after the UK (10.2 GW).

Table 1.2 Stages and key policies of offshore wind power in China

Stage	Period	Policy	Main content or sign
Demonstration project	Before 2010	"The Renewable Energy Act" of 2006 In 2009, the National Energy Administration issued the "Offshore Wind Farm Project Planning Work Outline"	The Shanghai Donghai Bridge Offshore Wind Farm became the first domestic large-scale offshore wind power project demonstration
Concession bidding	2010–2014	In 2010, the National Energy Administration and the Oceanic Administration issued the "Interim Measures for the Management of Offshore Wind Power Development and Construction"	From May to September 2010, the National Energy Organization invited bids for the first round of domestic offshore wind power concession projects
Fixed feed-in tariff	2014–2018	In June 2014, the National Development and Reform Commission issued the "Notice on Offshore Wind Power On-grid Tariff Policy"	It stipulates that the on-grid electricity price of intertidal offshore wind power and offshore wind power projects put into operation in 2017 is 0.75 yuan/kWh and 0.85 yuan/kWh respectively
Competing configuration	From 2019	In May 2018, the National Development and Reform Commission issued the "Notice on Relevant Requirements for Wind Power Construction in 2018" and "Guidance Plan for Competitive Allocation of Wind Power Projects In May 2019, the National Development and Reform Commission issued thc "Notice on Improving Wind Power On-grid Tariff Policy"	From 2019, all newly approved offshore wind power projects should be allocated and determined through competition Both onshore and offshore wind power benchmark electricity prices will be changed to guide prices, and the on-grid electricity prices determined through competition for newly approved projects in resource areas shall not be higher than the guide prices

The cumulative installed capacity of offshore wind power in China from 2011 to 2020 is shown in Fig. 1.4.

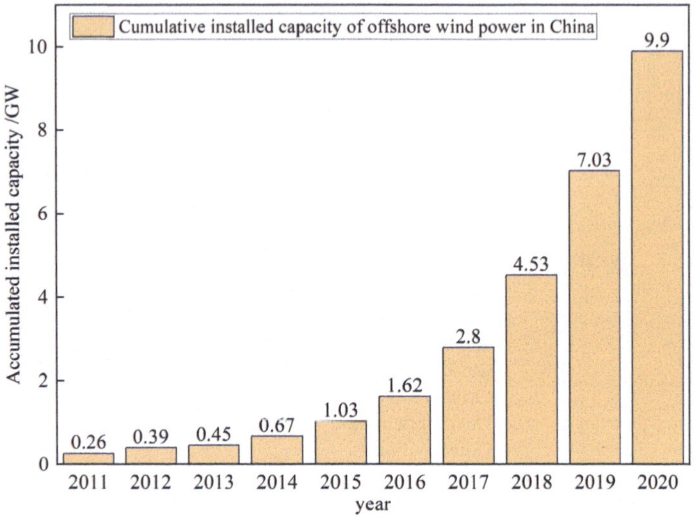

Fig. 1.4 Cumulative installed capacity of offshore wind power in China from 2011 to 2020 (*Source* GWEC, CWEA, National Energy Administration)

1.4 Major Issues and Development Trends of Offshore Wind Power

1.4.1 Major Issues with Offshore Wind

In recent years, the global cost of offshore wind power has dropped sharply. It is estimated that by 2030, the global installed capacity of offshore wind power will increase by about 5 times on the current basis. The rapid development of offshore wind power will inevitably highlight existing problems and bring some new ones.

(1) Compared with onshore wind power, harsh marine environments such as high humidity, high salt spray, long sunshine, seawater, and ice floes will inevitably make offshore wind power face technical challenges regarding corrosion protection, operation, maintenance, etc.

Offshore wind turbines, from the unit, the tower to the underwater foundation, are located in areas such as ocean atmosphere, spray splash, tidal range, full immersion, and sea mud. The corrosion environment and corrosion characteristics of each area are different. Many factors corrode offshore wind turbines, and it is particularly important to choose a reasonable anti-corrosion technology to improve the operating efficiency and quality of the equipment.

The operation and maintenance of offshore wind power are complex: on the one hand, due to the influence of the special environments (such as the influence of high humidity and high salt fog on equipment and the influence of weather factors on the maintenance window period), the reliability of equipment is poor, the failure rate is high, the maintenance cycle is long, the maintenance process is complicated; on the other hand, it is also affected by factors such as the reliability of wind turbines has not been fully verified, the professionalism of the operation and maintenance team needs to be improved, and the remote fault diagnosis and early warning capabilities are not perfect.

(2) The global offshore engineering construction capacity has not kept pace with the rapid development of offshore wind power, and the shortage of offshore wind power installation ships continues to plague the global offshore wind power industry.

As offshore wind farms gradually move toward deep seas and far seas, offshore projects' logistics, transportation, and installation processes have become increasingly complex, and the demand for offshore wind power installation ships with larger capacity wind turbines and more sophisticated technology has gradually increased. However, the development situation of the global shipping industry is still unable to meet the above requirements. At present, there are a total of about 60 offshore wind power installation ships available in the world, most of which are operating in the North Sea of Europe and some waters of China, which makes it difficult to meet the needs of countries around the world. A new report by consulting firm Rystad Energy shows that with the rapid development of offshore wind power, there will be a massive shortage of offshore wind power installation ships after 2025.

(3) As countries pay more and more attention to environmental and ecological protection, the impact of offshore wind power development on the marine environment and marine life is highly valued.

First of all, in the construction process of offshore wind power projects, wind turbines and power transmission cables need to be erected. The former needs to be directly driven into the seabed by piling, and the latter may also need to dig deep trenches during the laying process, which will stir up seabed sediments and cause sediment suspended. In addition, during the construction process, construction materials may accidentally leak, polluting the water quality of the nearby sea area and affecting marine life to a certain extent, which is not conducive to the balance of the marine environment. Secondly, for the organisms that reproduce and migrate in this sea area, wind turbines' noise and electromagnetic fields may affect the normal survival and reproduction of related organisms. In the European market, there have been cases where offshore wind power projects were suspended due to site selection that affected seabird activities.

(4) Improving equipment reliability, improving equipment localization, and reducing costs are the main issues facing the development of offshore wind power in China in the future.

The commercial operation of offshore wind power in China is relatively short, and domestic offshore generating units face obvious shortcomings such as a lack of effective verification of technology and a lack of standards, and there is a clear gap with experienced countries in Europe. Currently, China's grid-connected and commercially operated offshore wind farms are still in the initial stage of operation, and quality problems frequently occur.

In the past two years, new large-capacity units have been operated in batches, and the reliability still needs time to test. Reliance on the import of crucial equipment, low localization rate, and high cost are also important factors restricting the development of offshore wind power in China. Domestic offshore wind turbines are generally launched from land-based wind turbines after adaptive transformation such as anti-corrosion or imported mature equipment from abroad, especially high-power offshore wind turbines. Due to the inability to achieve localization, limited by the scale of production and technical level, the cost of domestic offshore wind turbines is relatively high, ranging from 5000 to 8000 yuan/kW.

1.4.2 Development Trend of Offshore Wind Power

In recent years, offshore wind power at home and abroad has generally shown a development trend of "from small to large, from near to far, from shallow to deep." It has made great progress in the field of technology integration and critical equipment for large-scale offshore wind power development. The Danish wind energy research and consulting agency predicts that driven by the accelerated construction of offshore wind power in China and the further mature development of the European market, the global offshore wind power industry will develop steadily from 2017 to 2026, with a compound average growth rate of 16%. The International Renewable Energy Agency (IRENA) believes that larger wind turbine blades and more complex floating platforms will extend the construction of offshore wind farms to the far sea to obtain more incredible wind energy and higher power generation in far sea areas. The development trend of offshore wind power can be summarized in the following aspects.

(1) The rated capacity of a single unit is gradually increasing, and offshore wind turbines have entered the era of 10 MW +. Since 2015, the rated capacity of wind turbines has been growing at a constant rate of 16%, and in 2020, the average rated installed capacity of wind turbines is 8.2 MW. At present, the offshore wind turbine with the largest single-unit capacity in the world is the Haliade-X model released by GE Renewable Energy of the United States, with a rated capacity of 13 ~ 14 MW. In 2020, orders for wind turbines have shown a trend towards next-generation sizes, with projects coming online after 2022 with a unit capacity between 10 and 13 MW.

(2) The scale of wind farms is getting bigger and bigger, and the single scale exceeds one million kilowatts, and the trend of large-scale development is evident. According to statistics from Wind Europe, the scale of offshore wind farms has grown steadily from 2015 to 2020, with an average wind farm capacity of 788 MW in 2020, an increase of 26% over 2019. At present, the world's largest offshore wind farm is the Dogger Bank project in the UK, with three phases,

each with an installed capacity of 1.2 GW, forming a super large wind farm of 3.6 GW.

(3) The offshore distance and water depth of wind farms continue to increase, exceeding 100 km and 100 m, respectively, and the trend of the deep sea is obvious. Wind power giants will compete to develop floating offshore wind power. According to Wind Europe, floating projects in Europe will exceed 7 GW in the next ten years. In addition, there are few shallow seas with a depth of less than 50 m in Asia, and floating wind power that does not rely heavily on water depth will develop rapidly, especially in Japan, South Korea, and China's Taiwan region. Floating offshore wind is one of the areas where Asia can compete with or surpass Europe.

(4) Online bidding has become the latest model for the development of offshore wind power, and the cost of offshore wind power has gradually decreased. The United Kingdom and France have gradually implemented offshore wind power bidding policies since 2015, the Netherlands and Denmark since 2016, and Germany since 2017. In 2018, the National Energy Administration of China issued the "Notice on Relevant Requirements for Wind Power Construction Management 2018". The notice pointed out that in 2018, "offshore wind power projects that have not determined the investment subject should all be allocated and determined through competition." In 2019, "all offshore wind power projects should be allocated and determined through competition." In the future, as long as the offshore wind power regulatory framework remains stable. As the offshore wind power industry continues to grow, the cost of offshore wind power will gradually decrease.

Chapter 2
Offshore Wind Resource Characteristics and Measurement Evaluation

2.1 Introduction

China has a long coastline and abundant offshore wind resources. Compared with onshore, due to the different properties of the underlying surface, and the influence of waves and tropical cyclones, offshore wind resources have unique characteristics. Mastering the characteristics of offshore wind resources is an important basis for offshore wind power development, operation, and maintenance.

This chapter first analyzes the characteristics and geographical distribution of offshore wind resources, then introduces the impact of tropical cyclones on China's offshore wind power, and finally expounds on the measurement parameters, measuring instruments, evaluation, and cutting-edge technologies of offshore wind resources.

2.2 Characteristics of Offshore Wind Resources

Compared with onshore wind resources, offshore wind resources have obvious advantages [1, 2]:

- The average wind speed is high, as shown in Fig. 2.1, because the wind power density and wind speed have a cubic relationship, so under the same installed capacity, the power generation of offshore wind farms is higher than that of onshore wind farms, the higher the wind speed, the more obvious the advantages;
- Sea surface roughness is small. When the wind passes through the rough surface or friction objects, the size and direction will change. Sea surface roughness is small, and the influence on the flow field is obviously lower than that onshore;

© The Author(s), under exclusive license to Springer Nature Singapore Pte Ltd. 2025
Y. Liu et al., *Offshore Wind Farm Technology*,
https://doi.org/10.1007/978-981-96-1889-7_2

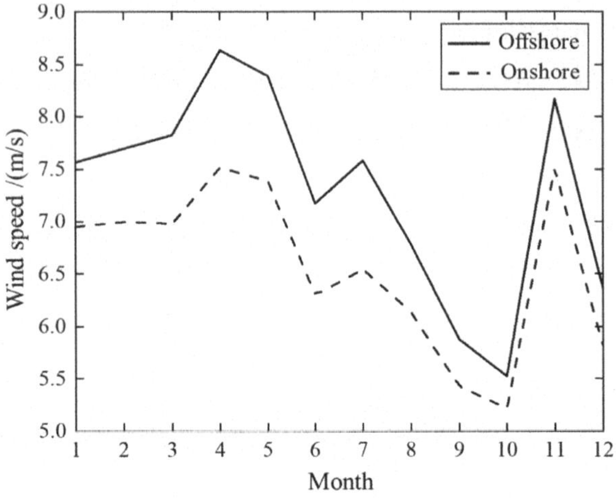

Fig. 2.1 Monthly comparison of offshore and onshore measured wind speed in a certain place (70 m height)

- Wind shear is low. Wind shear refers to the change of wind speed with height. The smaller the wind shear, the smaller the difference in wind speed at different heights. Even a lower hub height can bring good economic benefits. Low wind shear will reduce the fatigue load of wind turbines and prolong their lifespan.

In addition, offshore wind resources also have the advantages of low turbulence intensity and stable wind direction.

2.2.1 Variation Characteristics of Wind with Height

The variation of wind speed with height is different due to the flatness of the ground, the roughness of the ground, and the temperature of the wind channel. Table 2.1 shows the roughness of different underlying surfaces. Studies have shown that because the sea surface roughness is smaller than that of land, the wind speed 10 km offshore is more than 25% higher than that onshore.

The wind shear exponent α is used to characterize wind speed variation with height. A large α means that the wind speed increases rapidly with height, and the

Table 2.1 Roughness of different underlying surfaces

Ground condition	Roughness	Ground condition	Roughness
Smooth ground, hard ground, ocean	0.10	Many trees, few buildings	0.22 ~ 0.24
Grassland	0.14	Forest, village	0.28 ~ 0.30
Urban flats (higher grass and very few trees)	0.16	City (with tall buildings)	0.40
Tall crops, fences, and few trees	0.20	–	–

Table 2.2 Statistics of onshore and offshore wind shear index at different heights in Yancheng City, Jiangsu Province

Height	Onshore wind shear index	Offshore wind shear index
70 m/10 m	0.215	0.093
70 m/30 m	0.198	0.121
70 m/50 m	0.169	0.091
50 m/10 m	0.224	0.094
50 m/30 m	0.218	0.141
30 m/10 m	0.227	0.072
Overall fitted value	0.218	0.095

wind speed gradient is large; a small α means that the wind speed increases slowly with height, and the wind speed gradient is small. Table 2.2 shows the onshore and offshore wind shear index at various heights. The data used are from the two wind measuring towers onshore and offshore in Yancheng City, Jiangsu Province. It can be seen that compared with the wind shear index at various heights onshore, the offshore wind shear index is significantly smaller. Usually, at the height where wind turbines are installed, the gradient of wind speed change is already minimal, so increasing the tower height to increase the capture of offshore wind energy is not as effective as onshore. A comparison of onshore and offshore wind profiles is shown in Fig. 2.2.

$$v_2 = v_1 \left(\frac{h_2}{h_1} \right)^{\alpha} \tag{2.1}$$

where

v_2 wind speed at height h_2 above the ground, m/s
v_1 wind speed at height h_1, m/s
α wind shear index.

2.2.2 Turbulent Characteristics of Wind

Turbulence intensity describes the instantaneous change of wind speed relative to its average value, calculated by dividing the standard deviation of wind speed by the average value of wind speed for a period of time (usually 10 min), as shown in Table 2.3. Normally, the offshore turbulence intensity is less than that onshore. Since the offshore turbulence intensity is lower than that onshore, the recovery of the disturbance caused by the rotation of wind turbines is slow, and the downstream wind turbines need a larger separation distance from the upstream wind turbines.

The offshore turbulence intensity initially decreases with the increase in wind speed, and then gradually increases due to the increase of wind speed and waves, as

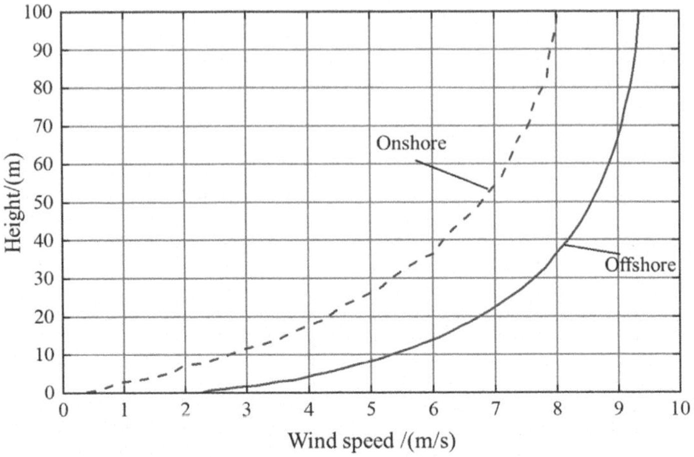

Fig. 2.2 Comparison of onshore and offshore wind profiles

Table 2.3 Offshore and onshore turbulence intensity at various heights (Yancheng City, Jiangsu Province)

Height	70 m	50 m	30 m	10 m
Onshore turbulence intensity	0.113	0.123	0.138	0.169
Offshore turbulence intensity	0.073	0.078	0.088	0.091
Differential relationship (onshore-offshore)	0.040	0.045	0.050	0.078
Scale relationship (onshore/offshore) (%)	64.60	63.40	63.80	53.80

shown in Fig. 2.3. In addition, the turbulence intensity decreases linearly with the increase in height, as shown in Fig. 2.4.

2.2.3 Distribution and Characteristics of Offshore Wind Energy Resources

1. **Distribution of global air pressure belts and wind belts**

The formation of global wind belts is closely related to the Distribution of air pressure belts [3]. Because the earth's surface receives uneven solar radiation, there are high and low air pressure distributions on the earth, which surround the earth's surface like several belts parallel to each other. Coupled with the differences in the nature of the underlying surface (including sea and land distribution, large terrain, surface friction, etc.), the global atmospheric circulation is formed, which constitutes the basic situation of global atmospheric operation and is also the source of global climate characteristics and weather conditions. From the equator to the poles, the

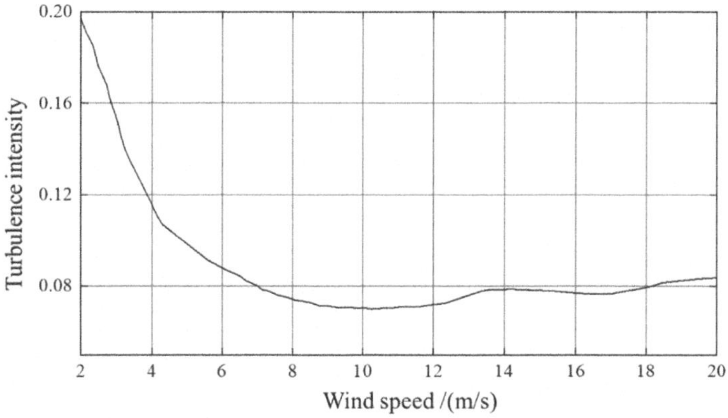

Fig. 2.3 Relationship between offshore wind speed and turbulence intensity

Fig. 2.4 Relationship between offshore turbulence intensity and height

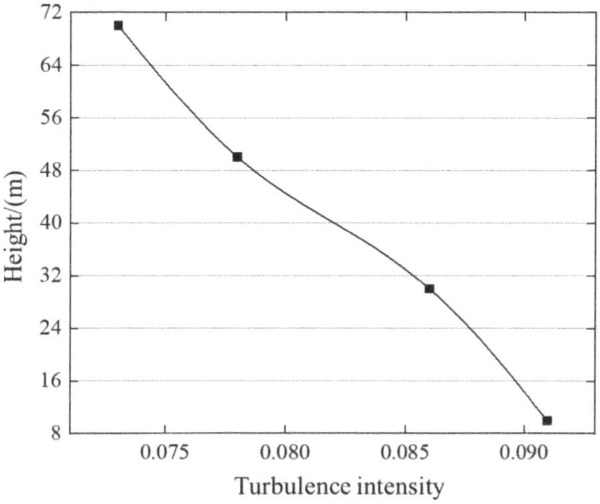

order of distribution of air pressure belts is equatorial low pressure, subtropical high pressure, subpolar low pressure, and polar high pressure. The air pressure varies from place to place, resulting in air pressure differences, causing air flow and forming corresponding wind belts. From the equator to the poles, the distribution of wind belts is the equatorial calm belt, the northeast (southeast) trade wind belt, the subtropical calms, the prevailing westerlies, and the polar easterlies.

2. **Distribution rules and characteristics of global offshore wind energy resources**

According to the global average wind speed onshore and along the coast, the area near the equator is basically the area with the lowest wind speed. The vicinity of the Tropic of Cancer belongs to the trade wind belt, and the wind speed is slightly higher.

The latitude of about 30 degrees in the northern and southern hemispheres belongs to the subtropical calms, and the wind speed is relatively small. Areas with higher latitudes belong to the prevailing westerlies, and the wind speed is generally very high, such as the North Sea region of Europe, where the wind speed is relatively high and the westerly wind prevails. The roaring forties between 40° and 60° latitude in the southern hemisphere have powerful westerly winds blowing all year round; the polar regions belong to the polar easterlies with relatively high wind speed.

In coastal areas worldwide, areas with extremely high wind energy density include the Atlantic coast of Europe and the coast of Iceland, the east and west coasts of the United States and Canada, and the coasts of Northeast Asia. Areas with high wind energy density include coastal areas of Southeast Asia and the South Asian subcontinent; Areas with low wind energy density are mainly near the equator. Generally speaking, the wind energy density in coastal areas is relatively high, which has excellent development value. The distribution of global offshore wind energy resources is shown in Table 2.4.

3. **Distribution of Offshore Wind Energy Resources in China**

China's coastline is about 18,000 km long, with over 6000 islands, a vast sea area, and abundant offshore wind energy resources. According to the census of wind energy resources by the China Meteorological Administration, the offshore wind power development potential in China is about 200 GW at a water depth of 5–25 m and a height of 50 m; At a water depth of 5–50 m and a height of 70 m, the offshore wind power development potential is about 500 GW, which has an excellent offshore wind

Table 2.4 Distribution of global offshore wind energy resources

Wind speed rating	Place
Extremely high wind speed (the wind speed is 8 ~ 9 m/s)	The southern coast of Africa
	The coast of Northeast Asia
	The islands' coast of the Caribbean
	The coasts of Australia and New Zealand
	The coasts of Chile and Argentina in South America
	The Atlantic coast of Europe and the coast of Iceland
	The eastern and western coasts of the United States and Canada and the southern coast of Greenland
High wind speed (the wind speed is 6 ~ 7 m/s)	The east coast of central South America, the coast of the South Asian subcontinent, and the coast of Southeast Asia
Low wind speed (wind speed below 5 m/s)	Continental coasts in the equatorial region: the west coast of Central America, the Atlantic coast of central Africa, and the coast of Indonesia

power development potential. Preliminary simulation results of China's offshore wind energy resources show that offshore wind energy resources are mainly concentrated in the southeast coast and nearby islands, with the most abundant wind energy resources in the Taiwan Strait, followed by eastern Guangdong, Zhejiang offshore, and central and northern Bohai Bay. Relatively speaking, areas with less offshore wind energy resources are distributed in the Beibu Gulf, the northwest, south, and southeast offshore waters of Hainan Island [4, 5].

At a height of 90 m in the Taiwan Strait, the annual average wind speed is basically between 7.5 and 10 m/s, and in some areas, the annual average wind speed can reach more than 10 m/s. This area is also one of the areas most affected by typhoons in China. From the Taiwan Strait to the south of Guangdong and Guangxi waters, at a height of 90 m, the annual average wind speed gradually drops to between 6.5 and 8.5 m/s. From the Taiwan Strait to the north of the Zhejiang, Shanghai, and Jiangsu waters, at a height of 90 m, the annual average wind speed gradually drops to between 7 and 8 m/s. Located in the Liaoning and Hebei waters around the Bohai Sea and the northern part of the Yellow Sea, at a height of 90 m, the annual average wind speed is basically between 6.5 and 8 m/s. The wind resources of coastal provinces (cities) in China are shown in Table 2.5.

China is rich in offshore wind energy resources, which mainly benefit from tropical cyclone activities in summer and autumn and the cold air in the north in winter and spring [6]. In winter, the northwest airflow between the Mongolian high-pressure and the Aleutian low-pressure drives directly southward, and the airflow deflects clockwise. The northwest wind prevails in the Bohai Sea and the Yellow Sea, the north or northeast wind prevails in the South Yellow Sea and the East China Sea, and the northeast wind prevails in the South China Sea. In spring, as the transition season from winter to summer, the wind direction in the Bohai Sea is chaotic, the northwest wind prevails in the Yellow Sea, and the northeast wind prevails in the East China Sea and the South China Sea; In summer, under the influence of the continental

Table 2.5 Wind resources of coastal provinces (cities) in China

Province (city)	Annual average wind speed (90 m, m/s)
Liaoning	6.5–7.3
Tianjin	6.9–7.5
Hebei	6.9–7.8
Shandong	6.7–7.5
Jiangsu	7.2–7.8
Shanghai	7.0–7.6
Zhejiang	7.0–8.0
Fujian	7.5–10
Guangdong	6.5–8.5
Guangxi	6.5–8.0
Hainan	6.5–9.5
Taiwan	7.5–10

low pressure, the Northwest Pacific subtropical high and the southwest monsoon, the southwest wind prevails off the coast of China; In autumn, the western Pacific subtropical high rapidly weakened and retreated southward, the Mongolian high-pressure and the Aleutian low-pressure reappeared, and the northerly wind prevailed off the coast of China.

2.2.4 Tropical Cyclone

A tropical cyclone is a low-pressure eddy that occurs over tropical or subtropical oceans. It is a powerful and deep tropical weather system that rotates counterclockwise in the northern hemisphere and clockwise in the southern hemisphere. A tropical cyclone that originates in the western Pacific, Northwest Pacific, and their adjacent waters is called Typhoon; in the Atlantic and eastern Pacific, it is called Hurricane; in the Indian Ocean and the South Pacific, it is called a Cyclonic Storm or simply Cyclone (Cyclone). China divides tropical cyclones in the Northwest Pacific Ocean and the South China Sea into six levels according to the maximum average wind speed near the bottom center, as shown in Table 2.6. In the table, those with a wind force rating of 12 or above are collectively referred to as typhoons. A typhoon is a circular cyclone with a diameter of about 1000 km. The central air pressure is extremely low. The typhoon eye is 10–30 km from the typhoon's center. The weather in the typhoon eye is calm, and the wind speed is very small. The weather on the outer wall of the typhoon eye is the worst, where the maximum destructive wind speed occurs. Outward from the wall of the typhoon eye is a spiral rain band; the further outward, the lower the wind speed.

The schematic diagram of typhoons is shown in Fig. 2.5.

Among the sea areas affected by tropical cyclones on a global scale, the Northwest Pacific Ocean has the highest frequency and strongest intensity of tropical cyclones, accounting for about 36% of the total. Therefore, among the countries developing offshore wind power, China is most affected by tropical cyclones, followed by the United States, Japan, Australia, and India, which will also be affected to varying degrees, and Europe is basically not affected.

Table 2.6 Classification of tropical cyclones in China

Tropical cyclone class	The maximum wind speed at the bottom center (m/s)	The maximum wind rating at the bottom center
Tropical depression (TD)	10.8 ~ 17.1	Level 6 ~ 7
Tropical storm (TS)	17.2 ~ 24.4	Level 8 ~ 9
Severe tropical storm (STS)	24.5 ~ 32.6	Level 10 ~ 11
Typhoon (TY)	32.7 ~ 41.4	Level 12 ~ 13
Severe typhoon (STY)	41.5 ~ 50.9	Level 14 ~ 15
Super typhoon (SuperTY)	$\geq$ 51.0	$\geq$ Level 16

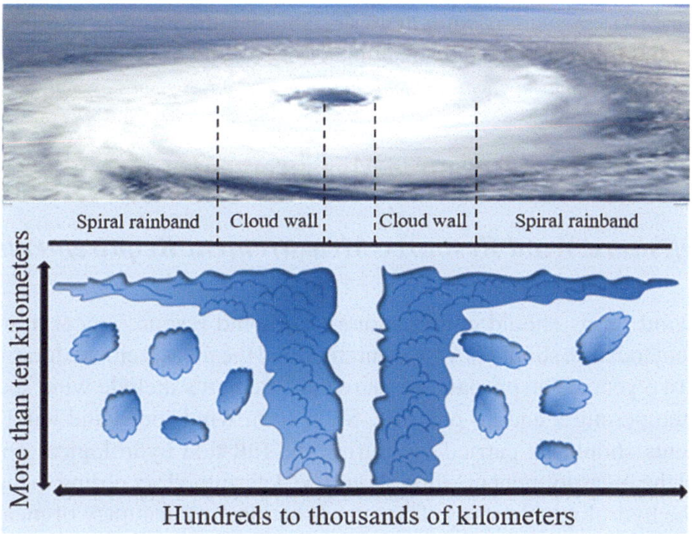

Fig. 2.5 Schematic diagram of a typhoon

China is the most affected by tropical cyclones among the countries developing offshore wind power in the world, the southeast coast is often affected by tropical cyclones every summer and autumn. On average, seven typhoons land in China every year [7]. Among them, Guangdong has the most, about 3.5 times, followed by Hainan (2.1 times), Taiwan 1.9 times, Fujian 1.6 times, Guangxi, Zhejiang, Shanghai, Jiangsu, Shandong, Tianjin, Liaoning, etc. a total of 1.7 times. It can be seen that the area affected by the typhoon decreases from south to north. In addition, typhoons usually affect the area from May to September.

The impact of tropical cyclones on wind power generation has two sides: on the one hand, tropical cyclones with less intensity and their peripheral circulation will bring about strong winds, and the wind speed is basically within the range of the cut-out wind speed (25 m/s) of the wind turbine. It can keep the wind farm in the "full power" state for a long time, thus bringing good power generation benefits. On the other hand, intense tropical cyclones, such as typhoons, will cause great damage to wind farms, including damage to wind farm equipment and roads, causing economic losses.

In 2006, the super typhoon "Sangmei" landed on the border of Fujian and Zhejiang, causing almost all the blades of nearby wind turbines to be destroyed. In 2018, super typhoon "Maria" landed in Fujian, causing two wind power equipment accidents: The No. 8 wind turbine in the Dajing Wind Farm collapsed, the No. 23 wind turbine blades broke, and the foundation was loose, the tower of No. 14 wind turbine in Luxia Wind Farm broke off. To reduce the impact of typhoons on offshore wind power, corresponding measures and countermeasures can be taken from various aspects such as equipment design, manufacturing, wind farm design, and operation to minimize the adverse impact of typhoons.

2.3 Offshore Wind Resource Measurement

2.3.1 Offshore Wind Resource Measurement Requirements

Offshore wind farms should conduct long-term wind resource measurements, the measurement location should be representative, and the measurement duration should be at least two years. The primary measurement elements include wind speed, wind direction, temperature, and air pressure. Short-term wind speed and wind direction measurements should be carried out during the full-tide hydrological observation period, and the measurement position should be determined according to the requirements of the hydrological survey. The time, location, and accuracy of measurement facilities that provide services for wind power prediction in wind farms should meet the relevant requirements for wind power prediction.

1. **Location and quantity of wind measuring points**

The location and quantity of wind measurement points should be determined in combination with the wind farm and its surrounding factors, according to the shape and scope of the wind farm site, considering the changes in wind resources in both directions parallel to and perpendicular to the coastline, and the measurement requirements during the operation phase [8]. They should be able to reflect the change of wind resources in the wind farm area. The location of the wind measurement point should avoid obstacles such as bridges, offshore drilling platforms, and islands, and the distance from obstacles should be greater than 30 times the height of obstacles. There should be at least one wind measuring point for a single wind farm. The control distance of wind measuring points of intertidal and subtidal mudflat wind farms on the vertical coastline should not exceed 5 km, and other offshore wind farms should not exceed 10 km.

Taking the mean sea level as the starting datum, the measuring height of the wind tower should be higher than the hub height of the pre-installed wind turbine, and there should be at least one wind measuring tower within the scope of the wind farm with a measuring height not less than 100 m. For the wind tower with a height of 100 m, anemometers should be installed at heights of 10, 50, 60, 70, 80, 90, and 100 m for wind speed measurement. Among them, two sets of anemometers should be installed at 10, 60, 80, and 100 m. Two sets of independent wind vanes are installed at 10, 60, 80, and 100 m of the wind tower. A barometer and a thermometer are respectively installed near the height of 10 m to measure air pressure and temperature. When conditions permit, a temperature sensor is installed synchronously on the top of the tower. When setting up the instrument at a height of 10 m, the influence of waves should be avoided, and it can be raised appropriately according to specific

Fig. 2.6 Schematic diagram of the offshore wind tower (*Source* OffshoreWind.cn)

conditions. When setting the measurement height of wind towers of other heights, it can be comprehensively determined concerning the hub height of the 100 m wind measuring tower and the pre-installed wind turbine.

The schematic diagram of the offshore wind tower is shown in Fig. 2.6.

2. **Measurement parameters and requirements**

The measurement parameters should comply with the relevant provisions of the current national standard "Methodology of wind energy resource measurement for wind farm" GB/T 18709 and the industry standard "Specification of wind energy resource measurement and marine hydrographic observation for offshore wind power projects" NB/T 31029:

- The sampling time interval of wind speed parameters should not be more than 3 s, automatically calculate and record the average wind speed every 10 min, the standard deviation of wind speed every 10 min, the maximum wind speed every 10 min, and its corresponding time and direction, the unit is m/s.
- The wind direction parameter sampling time interval should be no more than 3 s. The wind direction value is automatically calculated and recorded every 10 min; the unit is the degree (°).

- The temperature parameters should be sampled and recorded every 10 min, and the unit is °C. When conditions permit, a temperature difference measuring device should be used to simultaneously measure the wind tower's temperature gradient.
- Air pressure parameters should be sampled and recorded every 10 min; the unit is kPa.

2.3.2 Offshore Wind Resource Measurement Technology

According to different offshore wind resource data sources, offshore wind resource measurement technology can be roughly divided into offshore station observation and offshore satellite remote sensing detection [9]. Among them, the meteorological station, buoy station, wind tower, and LiDAR belong to station observations, which are not affected by weather and sea conditions and can continuously monitor wind conditions hour by hour. They are valuable site data sources for offshore wind resource assessment. Spaceborne microwave scatterometer, radiometer, altimeter, and multi-aperture radar belong to satellite remote sensing detection, invert the sea surface wind conditions based on the backscattering signals of gravity capillary waves on the sea surface, which is an effective means to obtain macroscopic sea surface wind conditions.

1. **Offshore station observation**

Conventional methods for offshore station observation mainly include laying buoys, building wind towers, or building weather stations on isolated small islands to monitor offshore wind conditions with high data accuracy continuously. Among them, buoys and weather stations are mostly single-layer observations, and the wind measurement height is generally about 10 m, the data are only suitable for statistical analysis of offshore wind farms near the sea surface. In the actual construction and development of offshore wind farms, the measurement data of the vertical wind profile is often needed to support the calculation and analysis of parameters such as hub height wind speed and atmospheric turbulence intensity, and then carry out a series of work such as wind turbine selection. Building wind towers and laying out multi-layer observations is a common way to obtain vertical wind profiles.

The No. 10 buoy in the East China Sea buoy array of the China Offshore Ocean Observation and Research Network is shown in Fig. 2.7.

There are many limitations to the measurement of offshore wind towers. In general, there are many measurement limiting factors such as tower shadow effect, measurement height, number of sensors, etc. In addition, there are high construction costs and high maintenance costs. Therefore, station wind measurement and telemetry technologies such as acoustic radar, microwave radar, and LiDAR have emerged properly. These new mobile wind measurement technologies provide a new solution to the above problems. At present, the most widely used technology is the LiDAR wind measurement technology.

Fig. 2.7 The No. 10 buoy in the East China Sea buoy array of the China offshore ocean observation and research network (*Source* Institute of Oceanography)

Lidar wind measurement technology uses lasers' Doppler frequency shift principle to obtain wind speed and direction information by measuring the frequency changes produced by light waves reflected by aerosol particles that encounter wind movement in the air. In this way, the corresponding height's vector wind speed and wind direction data can be calculated, and the effective wind measurement height can reach 200 m or even higher. Compared with the offshore wind towers, the LiDAR wind measuring height is higher, and the cost is lower (only 40 ~ 60% of the construction cost of the offshore wind towers). Moreover, it is more portable and mobile and has gradually become a new trend in offshore station observation. LiDAR wind measurement has been added to the revised IEC6 1400-24.

Offshore LiDAR wind measurement equipment is shown in Fig. 2.8.

2. **Offshore satellite remote sensing detection**

Fig. 2.8 Offshore LiDAR wind measurement equipment

Traditional offshore wind farm observation methods (buoys, weather stations, and wind towers, etc.) have obvious limitations, such as high construction and maintenance costs, inability to obtain large-area sea surface information, and more significant influence by external environmental factors, which affect the acquisition of high-quality data. In the 1960s and 1970s, with the development of space remote sensing technology, various spaceborne sensors emerged at the proper moment. Spaceborne sensors have the advantages of comprehensive spatial coverage and continuous time. The acquisition of satellite remote sensing detection data and the continuous advancement of satellite data wind field retrieval research has made up for the lack for offshore observation data. In the past 10 years, satellite remote sensing data have been widely used in wind resource assessment.

At present, the spaceborne sensors used for offshore wind farm detection mainly include microwave scatterometers, microwave radiometers, microwave altimeters, synthetic aperture radars, etc. [10, 11]. According to the working mode, spaceborne sensors can be divided into two types: active and passive: microwave radiometer belongs to the passive type, microwave scatterometers, microwave altimeters, and synthetic aperture radars belong to the active type.

Microwave scatterometers can measure sea surface wind speed, sea surface wind direction, etc. Microwave radiometers can measure ocean temperature, sea surface wind speed, seawater salinity, offshore oil pollution, sea ice thickness, area, icebergs, ice age, etc. Microwave altimeters can measure sea surface wind speed, sea surface height, wave height, current direction, current velocity, wave direction, etc. Synthetic aperture radar can measure ocean currents and waves, draw sea height maps, sea ice direction maps, and offshore oil pollution. The main information comparison of spaceborne sensors is shown in Table 2.7.

1. Microwave scatterometer

The microwave scatterometer indirectly obtains the sea surface wind field by measuring the sea surface's backscatter coefficient. Microwave scatterometers mounted on weather satellites measure sea surface roughness by emitting microwave

Table 2.7 The main information comparison of space-borne sensors

Spaceborne sensor	Measurement	Time resolution/h	Spatial resolution/km	Measuring range/(m/s)
Microwave scatterometer	Backscatter cross-section	48 ~ 72	25 ~ 50	4 ~ 26
Microwave radiometer	Radiant power	24 ~ 48	25 ~ 50	2 ~ 50
Microwave altimeter	Pulse delay and waveform, backscatter cross-section	240	6.7	2 ~ 18
Synthetic aperture radar	Backscatter cross-section	24	0.5 ~ 1	12.5 ~ 40

pulses and detecting backscatter back to the satellite, according to the functional relationship with the wind field, the wind field information at a height of 10 m above the sea (neutral condition) is obtained by retrieval.

2. Microwave Radiometer

The algorithms for retrieval of sea surface wind fields by microwave radiometers can be divided into three principal categories: statistical algorithm, semi-statistical algorithm, and physical algorithm. Statistical algorithms deduce empirical relational formulas through microwave radiometer brightness temperature and on-site measurement of ground object parameters, such as multiple linear regression or improved D matrix algorithm. The semi-statistical algorithm, also known as the physical statistical method, differs from the statistical algorithm in that the radiative transfer model simulates the brightness temperature during regression. The essence of the physical retrieval algorithm is to solve the nonlinear equations to minimize the difference between the SSM/I measurement results and the simulated brightness temperature.

Microwave radiometers and microwave scatterometers can complete a global earth observation in one day and quickly obtain a wide range of sea surface wind fields. Inverted wind field data are widely used for wind resource assessment in macroscopic sea areas, and they are often used as the basic data for numerical simulation of wind field forecasts or assimilation data for reanalysis of wind field data.

3. Microwave altimeter

The microwave altimeter takes the sea surface as the telemetry target, transmits the radar pulse signal to the sub-satellite point, obtains the echo waveform, and uses the waveform to determine physical quantities such as sea surface height, effective wave height, and backscatter coefficient. The backscatter coefficient and other parameters can also retrieve the sea surface wind speed.

The advantage of using a microwave altimeter to detect sea surface wind speed is that the along-track resolution (about 6 km) is much higher than that of a microwave scatterometer (about 25 km), and the accuracy (1.7 m/s) is also higher than that of microwave scatterometer (2.0 m/s). The disadvantage is that it can only detect sub-satellite points, and the repetition period is long (10 d and above).

4. Synthetic aperture radar

The brightness change effect of the image and the empirical geophysical model function (GMF) can be used to invert the sea surface wind speed from the original SAR image. The European Center for Medium-Range Weather Forecasts established the CMOD4 model function based on the C-band VV polarimetric scatterometer on ERS-1 and the numerically simulated wind farm data, which is the most widely used geophysical model function in SAR wind farm retrieval.

Synthetic aperture radar has the highest spatial resolution, is unaffected by clouds, sunshine, and sunlight, and can realize all-day and all-weather imaging. It has gradually become a new method for monitoring offshore wind farms, especially retrieval of nearshore wind farms.

2.4 Ocean Engineering Environmental Factors and Observation

In addition to the need to measure wind resources, offshore wind energy resource assessment should include marine hydrographic surveys and marine geological surveys, which require a comprehensive survey of ocean currents, sea ice, waves, and submarine geological structures. The basic concepts, main observation methods, and observation requirements of ocean currents, sea ice, and waves are introduced below.

2.4.1 Ocean Currents

Offshore ocean currents are usually divided into tidal and non-tidal currents. The tidal current is the periodic horizontal movement of seawater caused by the tidal force of celestial bodies. Non-tidal currents can be divided into permanent ocean currents and temporary ocean currents. Permanent ocean currents include ocean circulation, geostrophic currents, etc. Temporary ocean currents are caused by changes in meteorological factors, such as wind currents, nearshore wave currents, and atmospheric pressure gradient currents.

1. **Composition and classification of ocean currents**

Ocean current is a vector. The direction of the ocean current refers to the direction of the flow, in degrees (°), where true north is 0 and measured clockwise. Velocity refers to the distance of seawater flow per unit time, in m/s or kn (knot) as the unit, lkn = 1.852 km/h.

The flow of seawater can be divided into periodic and non-periodic, depending on the reasons for the formation of ocean currents. There are three main reasons for the formation of ocean currents that are closely related to ocean engineering: The first is the tidal current caused by tidal phenomena, which is periodic; The second is that the wind force acting on the vast sea surface causes the sea surface to drift, also known as wind current; The third is the density flow formed by the uneven distribution of seawater temperature, salinity, and even density due to the heating or cooling of the vast sea surface, evaporation or uneven precipitation. The following mainly introduces the tidal current and drift current, which are closely related to coastal engineering.

The tidal current corresponds to the tide, and there are semi-diurnal tides, diurnal tides, and mixed tides. Due to the difference between the seabed topography and the shape of the coast, tidal current phenomena are more complicated than tidal phenomena. Seawater flow during rising (falling) is called the rising (falling) tide. The tidal current is not only cyclical in flow velocity but also cyclical in the flow direction. According to the flow direction, there are two flow directions of the tidal current, which are rotating flow and reciprocating flow.

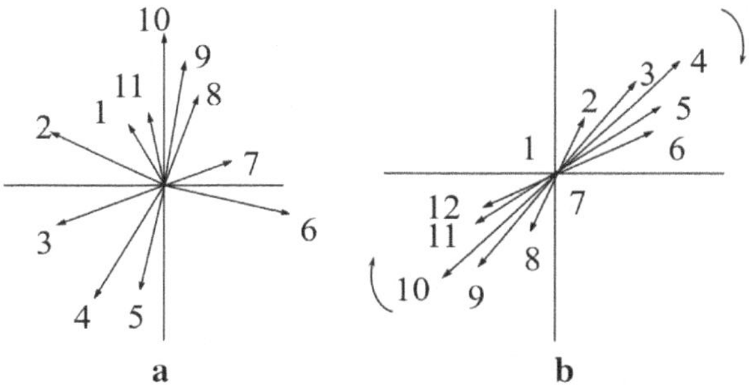

Fig. 2.9 **a** Rotating flow, **b** reciprocating flow

Rotating flow generally occurs in open seas and wide sea areas and is a common form of the tidal current. Due to the influence of the earth's rotation and seabed friction, the tidal current often does not simply form a reciprocating flow, but the flow direction changes continuously, as shown in Fig. 2.9a. Reciprocating flow often occurs in narrow straits, waterways, harbors, estuaries, and sea areas with many islands near the coast. Due to the terrain limitation, the tidal current mainly changes in two opposite directions, thus forming the reciprocating flow of seawater, as shown in Fig. 2.9b.

Drift current is caused by the friction between the wind and the sea surface, and the flow direction is affected by the inertial force of the earth's rotation. In the northern hemisphere, it is biased to the right of the wind direction, and in the southern hemisphere, it is biased to the left of the wind direction. Seawater surface friction makes the energy of surface seawater movement gradually transfer to the deep layer.

V.W. Ekman proposed the drift current theory in the early twentieth century to study drift current. The two assumptions of this theory are: drift current occurs in the vast ocean or the sea far away from the shore; the ocean's depth is infinite, at least enough so that the stable wind direction and wind force can cause constant drifting. The following conclusions mainly summarize Ekman drift current theory:

(1) The direction of surface drift current is 45° to the right of the wind direction in the northern hemisphere and 45° to the left in the southern hemisphere. This deflection does not change with changes in wind speed, current speed, and latitude.
(2) The empirical relationship between surface velocity and wind velocity is

$$V_0 = \frac{0.0127}{\sqrt{\sin\varphi}} U \tag{2.2}$$

where

U wind speed, m/s
φ latitude.

In the vertical direction, the drift current velocity decreases rapidly with the increase of water depth. When the water depth is infinite, the complex expression of the drift current velocity V at the depth z is

$$V = V_0 \exp\left[-\pi \frac{z}{D} - i\left(\frac{\pi}{4} - \pi \frac{z}{D}\right)\right] \quad (2.3)$$

where

D Friction depth, m

2. **Ocean Current Observation**

The main observation elements of ocean currents are flow velocity and flow direction, and the auxiliary quantities that need to be observed are water depth, short-term wind speed, and wind direction to understand the characteristics of the marine environment in more detail and more realistically. The main observation methods include ship anchor current measurement, anchor submerged buoy current measurement, and anchor open buoy current measurement. Commonly used current meters mainly include direct reading current meters, Andra current meters, and acoustic Doppler current meters.

Ocean current observation data should be processed with special software: firstly, the rationality of the originally collected data should be checked. Then, calculate the current velocity at each observation level, store the processing results in the data file according to the specified format, and correct the flow direction according to the technical performance of the instrument and the magnetic declination of the survey area. Finally, draw the time series vector and vertical distribution diagrams of the ocean current.

2.4.2 Sea Ice

Sea ice is the product of seawater gradually changing from liquid to solid after the temperature drops below freezing point in cold seasons, accompanied by salt precipitation in seawater. The destructive effect of sea ice is enormous. In heavy ice years, sea ice can block bays and waterways, destroy passing ships, and collapse marine structures, posing severe marine disasters. The force of sea ice on marine engineering structures is one of the main control loads for engineering design in cold sea areas. The sea ice forces on structures depend on the size and structural form of structures and are closely related to the physical and mechanical properties of sea ice.

1. **Structure and type of sea ice**

Sea ice is a general term for all ice that occurs at sea, including ice directly frozen by seawater and freshwater ice originating from land (river ice, lake ice, glacier ice, etc.). Sea ice is generally composed of solid water (pure water), a variety of solid salts, and brine pockets whose concentration is greater than that of primary seawater and trapped in the voids of the ice structure. During the formation of pure ice, salts in seawater are separated and transferred below. Brine pockets are the main reason sea ice is weaker than freshwater ice at the same temperature. Although ice is a crystalline material, the shape and size of individual ice crystals vary widely. They may be a flake, plate, or columnar, ranging from 1 mm to several centimeters. When the ice lattice is ordered, the deformation and strength of the ice are generally anisotropic.

Ice is a hexagonal crystal composed of hydrogen and oxygen ions. The crystal-to-crystal arrangement changes as the freezing process progresses. Sea ice undergoes various stages of change during its formation process, presenting different state characteristics. In different icing periods, sea ice has different icing characteristics. According to different classification bases and standards, sea ice can be divided into various types. The commonly used sea ice classification is based on the growth process, surface characteristics, crystal structure, movement form, density and melting process, etc. The types of sea ice included in different classifications are shown in Table 2.8.

2. **Sea ice observation**

Sea ice observations mainly include floating ice observation and fixed ice observation. The main elements of floating ice observation are ice volume, density, ice type, surface characteristics, ice shape, drift direction and speed, ice thickness, and ice edge line. The main elements of fixed ice observation are ice type, thickness, and boundary. The auxiliary observation elements of sea ice are sea surface visibility, air temperature, wind speed, wind direction, and weather phenomena.

The sea ice observation should be carried out on the survey ship, and the observation will start immediately after the ship arrives at the station. The location of sea ice observation on the ship should be as high as possible, and the observation objects should be mainly sea ice beyond 2 times the length of the ship, to avoid the influence of the ship on sea ice observation.

Table 2.8 The types of sea ice included in different classification bases

Classification basis	The types of sea ice
Growth process	New ice, nilas, ice skin, pancake ice, grey ice, grey-white ice, white ice
Surface characteristic	Level ace, rafted ice, hummocked ice, pingo, iceberg, bare ice, ice cap
Crystal structure	Primary ice, secondary ice, superimposed ice, agglomerate ice
Movement form	Large ice sheet, medium ice sheet, small ice sheet, pack ice zone, pack ice, drift ice and ice tongue afloat, etc.
Density	Tight ice floe, very dense ice floe, dense ice floe, sparse ice floe, very sparse ice floe, ice-free zone
Melting process	Water pit ice, water hole ice, dry ice, rotten ice, flooded ice

2.4.3 Wave

Wave is one of the forms of seawater movement, resulting from seawater under the joint action of external force, gravity, and seawater surface tension. The periods and amplitudes of waves caused by different external factors are different. In 1965, Kinsman divided the types of waves according to the wave cycle, combined with the main disturbance force and restoration force, and gave an approximate energy distribution, as shown in Fig. 2.10. The wave with the shortest period is the capillary wave. Earthquakes, tsunamis, or storms cause long-period waves with a period greater than 5 min. The most notable energy distribution is the gravity wave, whose period is in the range of 1–30 s, especially in the range of 4–16 s. It occupies an important position in marine engineering and is the main wave load that marine structures need to consider.

Gravity waves are caused by wind; it is an umbrella term for wind waves, swells, and nearshore waves. Wind waves mainly refer to waves generated by the direct action of the wind. Swells are waves that travel into calm waters when the wind stops, turns, or leaves the wind area. The swells propagate into the shallow water area and are deformed due to the influence of water depth and terrain changes; refraction, diffraction, and breaking of waves occur, forming nearshore waves.

Figure 2.11 shows the wave's change curve with time. The main observation elements of waves are wave height, wave period, wave direction, waveform, and sea state, and the auxiliary elements are wind speed and wind direction. Acoustic wave meters, gravity wave meters, and pressure wave meters with wave direction are usually used.

Wave records are given in the form of simulated curves. Measure the vertical distance between a significant peak and a significant trough between two adjacent up

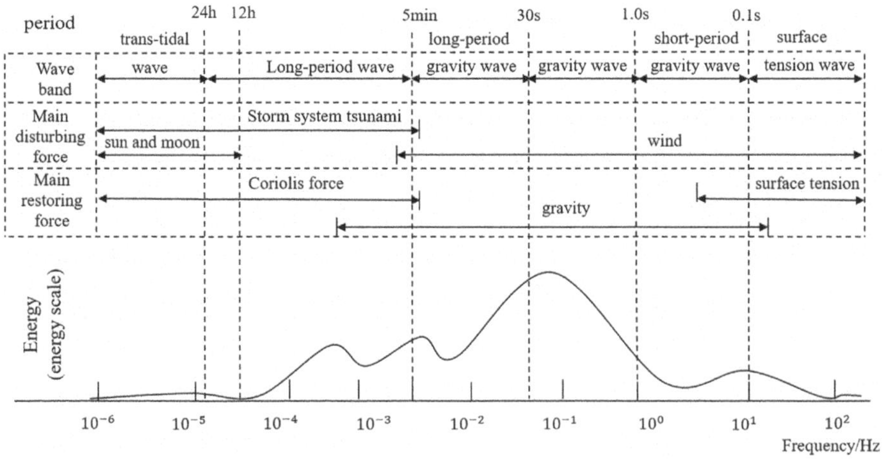

Fig. 2.10 Wave period, energy, and type

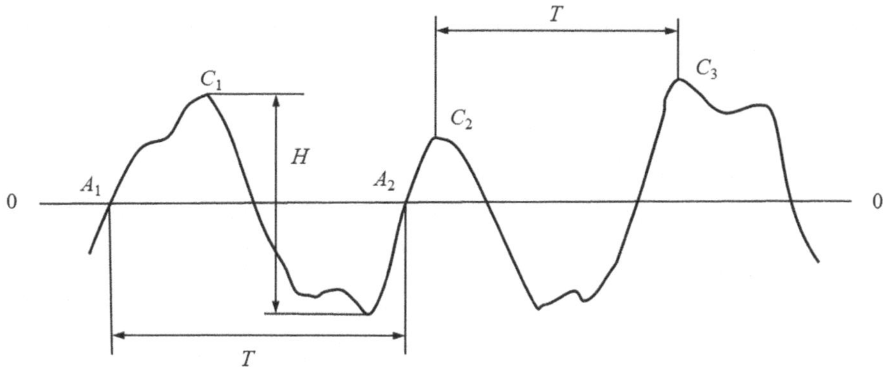

Fig. 2.11 The change curve of the wave with time

Table 2.9 Observation methods and time requirements for each marine hydrological element

Hydrological element	Station type	Observation method	Time requirement
Ocean current	Ocean current long-term station	Continuous observation	Not less than 1 year and synchronized with offshore wind farm measurements
Wave	Wave long-term station	Continuous observation	Not less than 1 year and synchronized with offshore wind farm measurements
Sea ice	Sea ice station	Large-scale observation	Ice period

span (or down span) zero points (A_1, A_2 in Fig. 2.11) in the wave continuous record as the wave height of a wave, namely H. . Measure the time interval between two adjacent significant wave peaks (C_2, C_3 in Fig. 2.11) or two up span zero points as a wave period T, and calculate the effective wave height and effective wave period according to the definition of effective wave period. The maximum value of all wave heights is selected as the maximum wave height, and the corresponding period is the maximum wave period. Table 2.9 shows each marine hydrological element's observation methods and time requirements.

2.5 Offshore Wind Resource Assessment

From the perspective of wind energy, the most notable characteristic of wind resources is variability. The wind is greatly affected by geographical environment and time factors. The cubic relationship between wind speed and available energy highlights the importance of wind resources [12]. Offshore wind resource assessment is the primary prerequisite for determining the development plan for offshore wind energy and completing the construction of offshore wind farms. The purpose of assessing wind resources is mainly to grasp the wind resources, provide a basis for determining the type of wind turbines and the micro-site selection of wind farms, and calculate the power generation of wind farms to facilitate the economic and technical evaluation of the entire project. This section first introduces wind measurement data processing and wind energy analysis and then introduces the relevant content of wind resource assessment.

2.5.1 General Steps of Offshore Wind Resource Assessment

1. **Wind measurement data processing**

To ensure the reliability of the assessment results of offshore wind resources, the integrity and rationality of the wind measurement data shall be tested first, and the test methods and reference values shall comply with the relevant provisions of the current national standard "Methodology of wind energy resource assessment for wind farm" GB/T 18710. Unreasonable and missing data should be handled in the following ways:

1. Take the data of other observation layers in the same tower as the missing data as the reference data. If the wind speed at other height layers of the same tower is also unreasonable, then the records of the same height layer at the observation point near the missing measurement point and with similar topographic features should be selected as reference data;
2. Analyze the impact of the tower shadow on two sets of wind speed data set at the same height of the wind measuring point;
3. The data during the freezing period should also be analyzed and processed, and should not be deleted directly.

2. **Calculation of wind resource eigenvalues**

According to the data processed in the first step, calculate the monthly and annual average wind speed, wind power density, and turbulence intensity of each wind measuring point at different heights. As well as the annual effective data integrity rate, wind shear index, air density, wind direction and wind energy distribution, Weibull distribution parameters, etc., the calculation method should comply with the current national standard "Methodology of wind energy resource assessment for wind farm" GB/T 18710.

3. **Analysis of Representative years**

After calculating wind resource eigenvalues, analyze the basic situation of long-term measuring stations around the wind farm and select representative reference measuring stations. According to the wind measurement data of the selected reference stations, the representativeness of the actual measurement period of the on-site wind measurement points in the long-term sequence is analyzed, and the multi-year representative correction is performed on the processed observation data of the representative wind measurement points of the wind farm.

When there is no long-term measuring station around the wind farm or the long-term measuring station is not representative, the reanalysis data can be used for multi-year representative analysis. The selected reanalysis data should be consistent with the daily, monthly, and annual wind speed trend corresponding to the corresponding wind measurement point and have a good correlation.

In the study of offshore wind resource assessment, global reanalysis data are mainly provided by the National Center for Environmental Prediction (NCEP), the European Center for Medium-Range Weather Forecasts (ECWMF), the National Aeronautics and Space Administration (NASA), and the Japan Meteorological Agency (JMA).

4. **Spatial representative analysis of wind resources**

According to the analysis results of the wind measuring points, combined with the location and scope of the wind farm and other factors, the representativeness of the wind measuring points to the wind resources of the wind farm is analyzed, simulate and calculate the distribution of wind resources in the wind farm area, and then analyze the changes of wind speed, wind direction, wind shear, turbulence intensity level, and wind power density with height at the wind measuring point.

5. **Wind resource assessment**

Using the revised data to calculate various parameters required for wind resource assessment of offshore wind farms, including hub height, annual and monthly average wind speed, wind power density, wind speed frequency distribution, wind energy frequency distribution, wind direction frequency, and wind energy density direction distribution, wind shear index, and turbulence intensity, etc., and draw various wind condition parameter charts. The calculation method should comply with the relevant provisions of the current national standard "Methodology of wind energy resource assessment for wind farm" GB/T 18710.

Wind power density is a comprehensive index for assessing wind farm resources, reflecting the effects of wind speed, the frequency distribution of wind speeds, and air density. The wind power density levels of offshore wind farms are shown in Table 2.10. When using, attention should be paid to the difference between the standard conditions on which the wind speed reference values in Table 2.10 are based and the actual conditions of the wind farm.

Table 2.10 The wind power density levels of offshore wind farms

Wind power density level	50 m height		70 m height		80 m height		90 m height		100 m height	
	Wind power density (W/ m2)	Annual average wind speed (m/s)	Wind power density (W/ m2)	Annual average wind speed (m/s)	Wind power density (W/ m2)	Annual average wind speed (m/s)	Wind power density (W/ m2)	Annual average wind speed (m/s)	Wind power density (W/ m2)	Annual average wind speed (m/s)
1	< 200	5.6	< 230	5.8	< 240	5.9	< 250	5.9	< 260	6.0
2	200 ~ 300	6.4	230 ~ 340	6.6	240 ~ 350	6.7	250 ~ 360	6.8	260 ~ 380	6.9
3	300 ~ 400	7.0	340 ~ 440	7.2	350 ~ 460	7.3	360 ~ 480	7.4	380 ~ 500	7.5
4	400 ~ 500	7.5	440 ~ 550	7.8	460 ~ 570	7.9	480 ~ 590	8.0	500 ~ 610	8.0
5	500 ~ 600	8.0	550 ~ 660	8.3	570 ~ 690	8.4	590 ~ 710	8.5	610 ~ 740	8.6
6	600 ~ 800	8.8	660 ~ 880	9.1	690 ~ 920	9.2	710 ~ 950	9.3	740 ~ 980	9.4
7	800 ~ 2000	11.9	880 ~ 2180	12.3	920 ~ 2270	12.5	950 ~ 2350	12.6	980 ~ 2430	12.8

Note 1. The reference value of the annual average wind speed at different heights is calculated based on the wind shear index of 0.10. 2. The annual average wind speed reference value corresponding to the upper limit of wind power density is calculated assuming that the sea level standard atmospheric pressure and wind speed frequency conform to the Rayleigh distribution

2.5.2 Uncertainty Analysis of Wind Resource Assessment

The reliability of offshore wind resource assessment does not entirely depend on the quality of observation data and numerical simulation techniques, and the uncertainty is also related to many factors. Firstly, the wind parameters contain various error sources in the calculation process. Secondly, in the numerical model simulation, the selection of initial boundary conditions, the parameterization of physical processes, the nesting method, and the resolution setting will all introduce sources of uncertainty and error.

Among many uncertain factors, the long-term change in wind speed is the main factor of uncertainty in wind resource assessment and the most unstable factor in the risk analysis of wind power investment. Although long-term changes in wind speed in the study area can be obtained using long-term observation data, the assessment results obtained from historical data cannot fully represent the future trend of wind resources.

Therefore, after completing the basic offshore wind resource assessment, it is necessary to analyze the influence of uncertainty on the results to judge their reliability. At present, the uncertainty analysis of offshore wind resource assessment includes wind speed measurement, representative year analysis, climate change, wind shear calculation, and wind resource spatial distribution simulation. In addition, the uncertainty factors of wind resource assessment should be analyzed to estimate the conclusions of wind resource assessment under different probabilities.

2.6 Frontier Technologies of Offshore Wind Resources

2.6.1 Numerical Simulation Technology for Offshore Wind Resource Assessment

Offshore wind power is gradually developing towards large-scale, far-reaching offshore, and affordable. In the face of broad development prospects, the technical challenges faced by offshore wind power development and construction should not be underestimated [13]. Compared with onshore wind power, the lack of offshore hydrometeorological observation data and the high cost and difficulty of offshore observation further increase the technical difficulty of offshore wind resource assessment. This section focuses on the current cutting-edge numerical simulation technologies for offshore wind resource assessment.

At present, the modeling methods mainly used in the wind power industry include linear models, Computational Fluid Dynamics (CFD) models, and mesoscale meteorological models. Compared with the linear model, the CFD model adopts a turbulence closure scheme that is more in line with the atmospheric boundary layer, can consider more atmospheric physical characteristics such as atmospheric stability, and has stronger scalability [14]. Existing models and software are mainly developed around the onshore environment, and their applicability in the development of offshore wind farms remains to be studied. Changes in offshore observation methods have also put forward higher requirements for the data assimilation process of wind energy assessment models and software. Therefore, research on resource assessment and planning design software suitable for marine environments is an essential direction for developing offshore wind power technology.

1. **Multi-scale nested model**

The wind field in nature has the characteristics of variation in multiple temporal and spatial scales, and the influence of wind field changes at different scales on wind farms is different. A research review published by the National Renewable Energy Laboratory in the top international journal Science in 2019 pointed out that improving the understanding of atmospheric physical processes and flow field characteristics of wind farms at different scales is one of the main challenges in wind energy research and development. Among them, the nested use of mesoscale and small-scale models can effectively solve the problem of insufficient applicability of a single-scale model, which is the development trend of numerical simulation technology for offshore wind resources [15]. Numerous studies have shown that mesoscale and small-scale nested models can simulate wind fields better than either model alone, especially near the coast. At present, although there are some research and applications on the wind resource assessment model based on mesoscale and small-scale nesting, it still takes some time before its large-scale promotion and application in the field of offshore wind power. How improve the simulation accuracy of mesoscale meteorological models, solve the fitting problem of different scale models, and improve the operating efficiency of nested models are the main problems to be solved in the future.

2. **Coupled air-sea model**

Traditional wind resource assessment modeling methods only regard the ocean as a type of underlying surface roughness of the model, ignoring the influence of air-sea interaction, which affects the accuracy of numerical simulation of offshore wind resource assessment. Existing studies have shown that the sea surface's wave characteristics and thermodynamic properties can significantly affect the wind profile and turbulence structure of the atmospheric boundary layer over the sea. In recent years, the coupled forecasting system, including multi-components such as the atmosphere, ocean, sea ice, and land surface, has become the mainstream numerical forecasting system of domestic and foreign atmospheric and ocean operational forecasting centers. The coupled air-sea model can effectively improve the accuracy of simulation and forecast of offshore wind farms, provide integrated marine hydrometeorological information, and serve the comprehensive assessment of offshore wind power resources and the environment. With the continuous improvement of computing power, the coupled air-sea model will play an increasingly important role in the field of offshore wind power resources. How to improve the accuracy and adaptability of the coupled air-sea model in offshore wind power and enrich the application scenarios is the focus of future research.

2.6.2 GNSS-R Ocean Wind Fields Retrieval Technology

The amount of data and calculations often limits the numerical simulations described above, and the research time range is generally short. With the development of remote sensing technology, research on wind field retrieval using satellite data has been continuously advanced, making up for the lack of offshore observation data and making it possible to use satellite remote sensing data to evaluate offshore wind resources [16].

Global Navigation Satellite System Reflections (GNSS-R) technology is a new remote sensing method that has emerged in recent years and was developed from the Global Navigation Satellite System (GNSS). GNSS members include the Global Positioning System (GPS) of the United States, the Global Navigation Satellite System (GLONASS) of Russia, the Galileo Satellite Navigation System (GALILEO) of the European Union, and the Beidou Navigation Satellite System (BDS) of China, as shown in Fig. 2.12

GNSS-R technology uses the reflected signal of the navigation satellite L-band signal as the remote sensing source and has all-day and all-weather remote sensing capabilities. In addition, the L-band frequency is less affected by rainfall, has a low cost, no additional transmitter is required, and the receiver is small in power, light in weight, and has a wide coverage. Compared with single sensors such as microwave scatterometer, microwave radiometer, microwave altimeter, and synthetic aperture radar to obtain sea surface wind field, GNSS-R technology has rich signal sources, which makes up for the shortcomings of traditional measurement methods. A new sea wind observation method is provided for sea surface wind field retrieval.

Fig. 2.12 GNSS composition

Traditional sea surface wind speed retrieval methods by GNSS-R technology mainly include the waveform matching method and empirical function method [17]. The waveform matching method uses the theoretical correlation power waveform and the measured correlation power waveform, and the least square method is used to match and retrieve the wind speed. The empirical function method mainly constructs the functional relationship between a single characteristic parameter and wind speed and applies this functional relationship to wind speed retrieval. Both methods have certain disadvantages: The waveform matching method requires the establishment of a large number of theoretical waveform databases, which requires a large amount of memory space, complex calculations, and cumbersome operations; The empirical function method often only uses a very small number of physical quantities related to wind speed, ignoring other factors related to wind speed, and loses certain retrieval accuracy.

In view of traditional methods' shortcomings, machine learning algorithms such as neural networks are often introduced in the research of cutting-edge technologies for correction, and GNSS-R sea surface wind speed and wind direction retrieval methods based on neural networks are proposed [18].

Existing research has verified that the neural network method has the characteristics of short modeling time, fast inversion speed, and high precision, which has promoted the application research of machine learning algorithms in the field of GNSS-R sea surface wind field remote sensing. In addition to sea surface altimetry and sea surface wind field, GNSS-R technology is also widely used in sea ice dynamic monitoring, marine oil spill detection, soil moisture detection, and other aspects.

References

1. Feng Y, He Y, Zhu Q et al (2016) Temporal and spatial characteristics of offshore wind resources. Journal of Tsinghua University (Science and Technology) 56(05): 522–529.
2. Fu P, Qin P F, Zhang J J (2014) Research of spatio-temporal characteristics of offshore wind resources. Journal of China Institute of Water Resources and Hydropower Research 12(02): 155–161.
3. Song J, Liu Y Q. (2011) Overview of Global Offshore Wind Energy Resources // Proceedings of the 2011 National Wind Power Engineering Information Network Annual Conference.
4. China Meteorological Administration (2014) Detailed Investigation and Evaluation Report on National Wind Energy Resources. Meteorological Press, Beijing, 3: 86-90, 108–122.
5. China Meteorological Administration Wind and Solar Energy Resource Assessment Center (2011) Detailed Investigation and Assessment of China's Wind Energy Resources. Wind Energy (8): 5.
6. Zhang X Z, Xu J W (2010) Distribution of Offshore Wind Energy Resources in China. Annual Conference of the Renewable Energy Power Generation Committee of the China Electrotechnical Society 2009–2010 and Wind Power Technology Exchange Conference, Beijing, 2010.
7. Yang X S (2012) Wind Power Generation Technology and Wind Farm Engineering. Chemical Industry Press, Beijing.
8. Wang W, Liu W (2016) Offshore Wind Measurement Tower Foundation Design. China Water Power Press, Beijing.

9. Li Z Q, Song L L, Ma H et al (2016) Review of Methodologies for Offshore Wind Resource Observation and Assessment. Advances in Earth Science 31(08): 800–810.
10. Dou F L, Shang J, Guo Y et al (2017) Satellite Remote Sensing of Sea Surface Winds: Technique Status and Application Progress. Advances in Meteorological Science and Technology 7(4): 6.
11. Lin M S, He X Q, Jia Y J et al (2019) Advances in marine satellite remote sensing technology in China. Haiyang Xuebao 41(10): 14.
12. Pan H P, Xu C, Xue F F et al (2019) Study on Wind Energy Parameter Model of Offshore Wind Farm and its Application. Acta Energiae Solaris Sinica 40(10): 2994–3001.
13. Yi K, Zhang Z L, Zhang H et al (2021) Technical Status and Development Trends of Numerical Modeling for Offshore Wind Resource Assessment. Distributed Energy 6(01): 1–6.
14. Guo Q Y (2020) Research on Offshore Wind Energy Resource Assessment Methods Based on Multi-source Data from Satellites and Ground Observations. Zhejiang University, Hangzhou.
15. Colmenar-Santos A, Perera-Perez J, Borge-Diez D et al (2016) Offshore wind energy: A review of the current status, challenges and future development in Spain. Renewable and Sustainable Energy Reviews 64: 1–18.
16. Li H, Xia Q, Yin C et al (2013) The Current Status of Research on GNSS-R Remote Sensing Technology in China and Future Development. Journal of Radars 64: 1–18.
17. Liu Y H, Xu R G, Niu X L (2021) Artificial Neural Network (ANN) based GNSS-R sea surface wind speed retrieval method. Information Technology and Informatization 06: 40–44.
18. Gao H, Bai Z G, Fan D D (2019) GNSS-R sea surface wind speed inversion based on BP neural network. Acta Aeronautica et Astronautica Sinica 40(12): 198–206.

Chapter 3
Offshore Wind Farm Design

3.1 Introduction

Offshore wind farm design is essential for a wind farm's safe and economical operation. Compared to onshore wind farms, the development of offshore wind farms involves a wide range of constraints, including the Maritime Administration, the Bureau of Oceanography, the national defense, the military, submarine cables, and fishermen. In addition to meteorological factors, marine hydrological conditions and other factors affecting offshore wind farms are also issues that must be considered when designing offshore wind farms. The design life of offshore wind farms is 25–30 years, and the operating environment is even worse. Therefore, the design of offshore wind farms is more complex.

This chapter will systematically discuss various aspects of offshore wind farm design, combining the design rules of onshore wind farms, considering the characteristics of offshore wind resources and multiple constraints. The main content includes macro and microsite selection of offshore wind farms, electrical system design, the basic design of wind turbines and booster stations, and technical and economic evaluation.

3.2 Macro Site Selection of Offshore Wind Farm

3.2.1 Basic Concept and Process of Macro Site Selection for Offshore Wind Farms

As one of the critical links of offshore wind power development, macro site selection is the process of determining the construction site, development value, and development strategy of offshore wind farms through a detailed investigation of key factors

© The Author(s), under exclusive license to Springer Nature Singapore Pte Ltd. 2025

Y. Liu et al., *Offshore Wind Farm Technology*,
https://doi.org/10.1007/978-981-96-1889-7_3

such as regional wind resources, natural environment, seabed geology, grid connection conditions, transportation, and constraints. The macro site selection results of offshore wind farms determine whether enterprises can obtain economic benefits, which is the key basis for investment decisions of offshore wind farms.

The macro site selection process for offshore wind farms includes development planning research, development agreement signing, macro site selection data collection, macro site selection site survey, macro site selection report preparation, and macro site selection report review.

(1) **Development planning research**

In the macro site selection of offshore wind farms, it is necessary to carefully study the national and regional offshore wind power development planning and sea use management opinions, continue to track the dynamic progress of offshore wind power planning and projects, investigate the distribution of offshore wind resources in detail, and initially draw up developable sea areas in areas with rich wind resources.

(2) **Signing process of the development agreement**

After the preliminary development of the sea area, actively communicate with the local government and the competent units, submit and develop the application, and strive for the commitment of resource allocation and development rights. After reaching the preliminary development intention with the local government, prepare the offshore wind farm development agreement (including the coordinates, scope, and installed capacity of the sea area to be developed, the location of the wind tower, the proposed commencement time, and the work to be coordinated by the local government, etc.), and submit it to the company's planning and development department and the competent government department for approval. The company and the local government signed the offshore wind farm development agreement.

(3) **Data collection of macro site selection**

Before macro site selection, it is necessary to collect relevant information about the proposed development sea area and use it as the basis for further work, mainly including marine functional zoning, meteorological and hydrological data, power system planning documents, submarine topographic map, surrounding sea channel planning data, surrounding port and harbor planning data, military facilities distribution map, marine natural reserves, mineral distribution map, fishery resource distribution map, aquaculture area distribution map, tourist protection areas, environmental sensitive points, surrounding urban planning data, submarine pipeline distribution map, surrounding cultural relics survey data, offshore wind farm planning data, etc.

(4) **Site investigation of macro site selection**

Before the site investigation of the macro site selection, the offshore wind farm site needs to be preliminarily arranged according to the relevant data collected, such as determining the site area, wind turbine unit layout, submarine cable route, etc. The owner shall invite the personnel of local relevant departments to the proposed site for a field survey.

Meteorological and hydrological investigation: By collecting and analyzing local meteorological data, observing the vegetation and topography of the adjacent coast, understanding offshore wind conditions from residents, and on-site artificial wind measurement, the wind resources of the site are preliminarily determined, the coordinates of the offshore wind tower are defined, and the data such as wave, tide and current are measured on site, which is used to calculate the hydrodynamic loads of underwater buildings such as the wind turbine foundations.

Bathymetric survey: The site of the offshore wind farm shall be comprehensively surveyed with a sonar meter, the water depth of the submarine cable route and its surrounding area shall be proposed, and the bathymetric map shall be drawn to provide the basis for the microscopic site selection and transmission line design. Small interval crisscross sonar detection shall be used for some critical areas, and significant interval sonar detection shall be used for remote areas.

Geological investigation: Collect data on the surface soil of the seabed at various locations of the site and conduct seabed drilling surveys to understand the geological conditions of the seabed. The drilling depth is generally 20–40 m. Shell drilling or spiral drilling is used for drilling. The drilling network is determined based on local specific conditions. For the drilling depth of 30–40 m, a jack-up rig can be used.

Troubleshooting subversive factors: Understand the adverse factors of project development, such as fishery resources near the site, marine nature reserves, submarine pipeline distribution map, essential channels in the surrounding waters, mineral distribution, tourism reserves, and military facilities distribution area.

Determination of development scale: Preliminary understanding of the available site area for development and determination of the development scale of offshore wind farms.

(5) **Preparation of Macro Site Selection Report**

After the site survey is completed, comprehensively analyze the meteorological and hydrological conditions, water depth conditions, seabed surface soil and geological conditions, inflammatory factors, and other construction conditions, eliminate alternative sites that do not meet the construction conditions, propose alternative sites for development, and prepare the macro site selection report of the offshore wind farm.

(6) **Review of macro site selection report**

After preparing the macro site selection report of the offshore wind farm, relevant experts are invited to review the report and finally determine the site of the offshore wind farm to be developed according to the review comments.

3.2.2 Basic Principles for Macro Site Selection of Offshore Wind Farms

The basic principles to be followed for macro site selection of offshore wind farms are as follows [1]:

- Abundant wind resources.
- Excellent marine hydrology, geology, access system, and transportation construction conditions.
- The principle of "double ten" shall be followed: when the offshore distance of the wind turbine is not less than 10 km, and the mudflat width is more than 10 km, the water depth of the sea area shall not be less than 10 m.
- Meeting the requirements of marine functional zoning is necessary. In the use and management of sea areas, non-functional sea use projects are generally encouraged to be compatible with developing marine functional zones. Adjustments should be made in case of conflict with marine functional zoning.
- It must be coordinated with urban, shoreline utilization, and mudflat planning.
- Meet the requirements of ecological environment protection and minimize the impact on birds, aquaculture, and nature reserves.
- Avoid main channels, anchorages, and restricted navigation areas to minimize the impact on the route.
- The protection scope of submarine pipelines, such as communication, power, and oil and gas, should be avoided.
- The scope of military facilities should be avoided.
- Focus on large-scale development and avoid decentralized access to power systems.

3.2.3 Factors Affecting the Macro Site Selection of Offshore Wind Farms

Many factors affect the macro site selection of offshore wind farms, such as social conditions, natural conditions, environmental conditions, and planning conditions.

1. **Social conditions**

Macro siting of offshore wind farms requires analysis and evaluation of grid connection conditions and transportation conditions.

(1) Grid-connected condition

According to the local power grid capacity, voltage level, grid structure, load characteristics, and construction planning, reasonably determine the construction scale and development timing of offshore wind farms, try to be close to the corresponding voltage level substation or grid, and reduce line losses and line construction investment.

(2) Transportation conditions

The construction of offshore wind farms should consider the specific conditions of existing ports, main roads, secondary roads, railways, and sea routes, and whether the existing ports meet the storage requirements of large equipment such as wind turbine components, booster stations, and supporting equipment, and whether the existing roads and sea routes meet the convenience and cost-effectiveness of transportation of large equipment, to avoid new road investment.

2. **Natural condition**

(1) Wind resource conditions

Wind resources are the primary factor to be considered in the site selection of offshore wind farms. The exploitable value of wind resources is evaluated based on the site's average wind speed, wind frequency, prevailing wind direction, wind power density, and annual available hours of wind energy. The offshore wind farm site should not only have good wind resources and stable wind power but also avoid typhoon activity areas, and should investigate the development scale, avoid decentralized access to the power system, and meet the investment return requirements.

(2) Meteorological and sea conditions

According to the characteristics of the principal climatic elements in the sea area, such as temperature, atmospheric humidity, wind, cloud, visibility, sea waves, sea fog, and atmospheric pressure, assess whether the standard and abnormal weather conditions at the site for many years, such as typhoon, storm surge, tsunami, and thunderstorm, meet the construction and operation standards of the offshore wind farm. Sea conditions include temperature, salinity, waves, currents, and ice. It is necessary to ensure good sea conditions as much as possible to reduce direct or indirect damage to the wind turbine generator, booster station, and foundation caused by adverse sea conditions.

(3) Ocean depth and geological conditions

Most countries classify the water depth of 15 m as shallow water, and the water depth of the built offshore wind farms is mostly less than 10 m, making it too shallow for transportation, and the water depth of 5–10 m is better. The National Energy Administration and the State Oceanic Administration jointly issued the Detailed Rules for the Implementation of the Interim Measures for the Management of Offshore Wind Power Development and Construction, which stipulate that the site selection of wind farms is mandatory, that is, "in principle, offshore wind farms should be arranged in the sea area where the offshore distance is not less than 10 km and the depth of the sea area is not less than 10 m when the mudflat width exceeds 10 km". Geological conditions mainly include seismic intensity and regional geology. The type and activity of earthquakes directly affect the design of offshore wind farms, which require particular infrastructure. The site's landform and the foundation soil's composition and characteristics are important factors affecting the foundation investment. The foundation soil is preferable to the sand foundation, and the muddy foundation will significantly increase the foundation work quantity.

3. **Environment condition**

(1) Hydrodynamic and sediment scouring and silting effects

The construction of offshore wind farm projects will generally cause changes in the average velocity of the nearby seawater, resulting in changes in the tidal current field near the project area, causing changes in the scouring and silting environment of the project area, especially the impact on the sediment scouring around the pile foundation of the offshore wind farm, forming a scouring pit, which is not conducive to the stability of the pile foundation. Therefore, the impact of hydrodynamic and sediment erosion and siltation changes after the project construction should be considered when selecting the site.

(2) Ecological environment impact

The construction of offshore wind farms will affect the aquaculture activities of some fishermen. The electromagnetic radiation of wind turbines will make marine organisms and fish lose their way. The excavation of a submarine cable trench and the piling of the wind turbine foundation will lead to the diffusion of suspended sediment, causing water pollution in some waters and phytoplankton death. During the construction of offshore wind farms, machine noise, light, and electromagnetic fields will affect birds' feeding, breeding, and migration. Therefore, the site selection must strictly comply with the Marine Environmental Protection Law of the People's Republic of China to minimize the impact on birds, fisheries, and natural reserves.

4. **Planning conditions**

(1) Avoid the sea use scope of existing projects

The site of the offshore wind farm should avoid using the sea for existing projects, such as ports, anchorages, waterways, customary routes, planned reclamation areas, military-restricted areas, marine nature reserves, mineral resources utilization areas, fisheries and aquaculture areas, tourism reserves, subsea pipelines (communications, electricity, oil, and gas, etc.), and other defined particular uses of the sea.

(2) Meet relevant national and regional planning requirements

The site of the offshore wind farm shall meet the requirements of marine functional zoning, regional (shoreline) development planning, urban construction planning, mudflat development and utilization planning, regional power grid construction planning, marine environment, and ecological protection planning, and meet the relevant requirements of the national marine authorities for the development of offshore wind farms.

3.3 Micro Site Selection of Offshore Wind Farm

3.3.1 Basic Concepts and Processes of Micro Site Selection for Offshore Wind Farms

The micro-site selection of offshore wind farms, i.e., the selection and determination of offshore wind turbine locations, is a comprehensive technical and economic comparison of wind turbine layout schemes from the perspectives of wind resources and construction conditions.

Based on excluding sensitive restrictive factors such as ports, anchorages, waterways, minerals, military restrictions, marine natural protection, fishery and aquaculture, tourism protection, submarine optical cables, submarine pipelines (communication, electricity, and oil and gas), noise, etc., a comprehensive comparison and full demonstration are made on the wind resource conditions, model applicability, construction conditions, and design schemes of power collection lines at the selected points. Finally, the recommended and alternative micro-site selection schemes are given to ensure that each site is free of inflammatory and sensitive constraints and that the micro-site selection scheme for offshore wind farms is feasible while achieving optimal economic benefits.

The micro-site selection of offshore wind farms is generally divided into three stages: preliminary scheme, site survey, and scheme finalization.

1. **Preliminary Scheme Stage**

The preliminary scheme stage mainly consists of designing the layout scheme for offshore wind turbines, routes and lifting platforms, and submarine cable routes.

(1) Layout scheme design of offshore wind turbine

First of all, a wind resource analysis is carried out based on the collected wind data from the wind tower, the same period, and long-term data from the meteorological station, and the wind resource evaluation conclusion is drawn (at least one year's representative annual data, including wind speed, wind power density, etc.). Then, according to the collected actual topographic map of the sea area of the offshore wind farm site (the scale is 1:2000, and the contour distance is not more than 2 m), the wind resource evaluation conclusion, functional department requirements, local regional special requirements, etc., complete the offshore wind turbine unit layout and estimate the power generation based on the mapping map. Finally, considering the construction conditions of offshore wind turbine sites, the preliminary scheme of the selected and alternative sites is determined.

(2) Route and lifting platform design

Based on the offshore wind turbine point layout, the preliminary scheme design of the route and lifting platform is carried out according to the scaled wind turbine model data and transportation specifications. The design principle of the route is to meet the transportation requirements of various components of the offshore wind turbine while considering the requirements of functional departments and local and regional special requirements. The design principle of the lifting platform is to meet the placement and lifting requirements of various offshore wind turbine components and ultimately form a preliminary plan for the route and lifting platform design.

(3) Submarine cable route design

In addition to special requirements such as functional department requirements, sea area usage restrictions, and avoiding existing engineering sea use, the design of submarine cable routes for offshore wind farms also needs to consider the impact of factors such as seabed geological structure, seabed depth, maximum wave level, and corrosive environment, determine the submarine cable route and landing point, and form a preliminary plan for the design of submarine cable routes.

2. **Site survey stage**

Before the site survey stage, it is necessary to prepare the proposed wind turbine point coordinate table, route and lifting platform schematic diagram, submarine cable route schematic diagram, and outdoor operation tools in the preliminary micro-site selection scheme stage and to organize the owner unit, design unit, and wind turbine supplier to participate in the site survey. The owner is mainly responsible for the review of sensitive factors, such as military restrictions, marine natural protection, fishery and aquaculture, submarine pipelines, and other related restrictive factors. The design unit is mainly responsible for reviewing the results of the preliminary scheme stage. The wind turbine supplier is primarily responsible for reviewing the transportation and construction conditions of offshore wind turbines.

The specific work contents in the field survey stage are as follows.

(1) Site investigation of microsite selection scheme

Investigate the proposed wind turbine point coordinate table, route and lifting platform schematic diagram, and submarine cable route schematic diagram in the preliminary micro-site selection stage one by one, confirm the engineering geological conditions, marine hydrological environment, and various limiting factors of the wind turbine point, route and lifting platform, and submarine cable route, evaluate the feasibility of the construction conditions, and determine whether the construction requirements are met.

(2) Adjustment and confirmation of the micro-site selection scheme

After the site survey, the wind turbine generator points that do not meet the construction conditions will be eliminated or replaced with alternative points that meet the requirements. Re-adjust and re-design the design schemes of the route, lifting platform, and submarine cable line that do not meet the construction conditions, and the design unit shall estimate the power generation and check the safety to form the final scheme of the microsite selection of the offshore wind farm, which must be signed and confirmed by the three parties.

3. **Scheme determination stage**

The scheme finalization stage mainly involves the report summary, submission for review, and completion of the closing of the final scheme.

(1) Complete the micro-site selection report

According to the results of the site survey stage and the safety load report of the wind turbine supplier, the preliminary plan is modified and improved to form a micro-site selection report. Next, send the recommended wind turbine location plan to the supplier so they can complete the microsite selection review report. The microsite selection report mainly includes the design scheme and economic analysis of the offshore wind turbine location, route and lifting platform, and submarine cable route.

(2) Complete micro site selection report review

The construction unit organizes the review of the micro-site selection report. The participating units include the design unit (mainly involving wind resources, routes, submarine cable routes, budget estimates, etc.), the wind turbine supplier (mainly involving wind resources, loads, transportation, hoisting, etc.), and the construction unit (primarily involving the review experts, members of the project department, etc.). Review experts and construction units of microsite selection reports review and provide opinions on design schemes for offshore wind turbine locations, routes, lifting platforms, and submarine cable routes.

(3) Completion of micro site selection closing work

The design unit shall complete modifying the micro-site selection report according to the review comments and submit the final report as the basis for the design work in the subsequent construction drawing stage.

3.3.2 Basic Principles for Micro-site Selection of Offshore Wind Farms

The basic principles to be followed for micro-site selection of offshore wind farms are as follows:

- The offshore wind turbine is arranged at a location with high wind power density.
- Offshore wind turbines shall be arranged centrally to avoid decentralized grid connection.
- To avoid significant changes in seabed geological conditions and water depth, try to maintain the consistency of foundation form and construction technology.
- Minimize the influence of wake between offshore wind turbine units. National industry standard NB/T10103-2008 Technical Code for Micro Site Selection of Wind Farm Engineering: Row spacing is the distance between adjacent wind turbines perpendicular to the main wind direction, and column spacing is between adjoining wind turbines parallel to the main wind direction. The row spacing of offshore wind turbines should not be less than seven times the diameter of wind

turbines, and the column spacing should not be less than three times the diameter of wind turbines. Row spacing and column spacing can be adjusted for offshore wind farms with different main wind energy directions.
- The micro-site selection of offshore wind farms should follow the principle of saving and intensive sea use, conform to or be compatible with marine functional zoning, and meet the requirements of resource development and utilization and ecological environment protection.
- The layout of offshore wind turbines should take into account wind resources, site utilization, power collection lines, transportation, construction and installation, project costs, and other factors. It should also improve power generation while meeting safety requirements.
- Avoid using the sea for existing projects such as ports, anchorages, waterways, minerals, military restrictions, marine natural protection, fishery and aquaculture, tourism protection, submarine optical cables, and submarine pipelines (communications, electricity, and oil and gas).
- The construction unit, design unit, and wind turbine supplier shall confirm the micro-site selection results of offshore wind farms.

3.3.3 Factors Affecting the Micro-site Selection of Offshore Wind Farms

The main factors affecting the micro-site selection of offshore wind farms are as follows:

1. **Wake effect between offshore wind turbines**

Unlike onshore wind farms, offshore wind farms have a flat, open sea surface free of obstacles, low roughness, and slight variation levels. Offshore wind farms have low turbulence and slow recovery after wind speed decay, requiring a more prolonged recovery distance. Therefore, the wake effect of offshore wind farms is more significant, manifested by the longer propagation distance of the wake, the existence of multiple wind turbines and multiple row wind turbines with superimposed wake effects, and the slow recovery of wind speed. Suppose the wake effect of offshore wind farms is too large. In that case, it will not only reduce the power generation capacity of offshore wind farms but also further increase the fatigue load of wind turbines and the failure probability of various components, thereby reducing the service life of wind turbines. According to the national industry standard NB/T10103-2018 Technical Code for Micro Site Selection of Wind Farm Engineering, the overall average wake loss of wind farms should be less than 8%, and the wake loss of single wind turbine generator sets should be controlled within 15% [2].

2. **Water depth and engineering geological impact**

The operating water depth of offshore wind farms affects not only the foundation structure, foundation construction methods, submarine cable laying processes, and plans of wind turbines but also is a significant factor in wind turbine equipment's transportation and lifting methods. When constructing a single offshore wind farm, substantial changes in water depth should be avoided, and consistency between the foundation construction method and the submarine cable construction process should be maintained as much as possible. Engineering geology includes the geological stability, stratum distribution, and mechanical properties of the engineering area in the sea area. It can reduce the operation and maintenance costs during the construction and operation of offshore wind farms.

3. **Impact of air routes and pipelines**

The microsite selection of offshore wind farms should avoid routes and pipelines, such as reserving fishing boat routes, island-to-island routes, and reserving a safe distance from submarine optical cables and pipelines.

4. **Other influencing factors**

In addition to avoiding prohibited construction areas such as ports, anchorages, waterways, minerals, military restrictions, marine natural protection, fishery and aquaculture, and tourism protection, other factors should also be considered to avoid, such as reefs, underwater obstacles, beacon lights, etc.

3.3.4 Selection Scheme for Offshore Wind Turbine Units

Wind turbine selection is one of the most critical steps in constructing offshore wind farms. The appropriate wind turbine can not only save the engineering investment of offshore wind farms but also improve the revenue of offshore wind farms and reduce the operation and maintenance costs. In addition to meeting the basic principles of onshore wind turbine selection, offshore wind turbine selection also needs to focus on the following influencing factors.

1. **Meteorological environmental conditions**

The meteorological environment conditions of offshore wind farms are significantly different from those of onshore wind farms. The selected wind turbine should adapt to complex environmental conditions at sea and perform better moisture resistance, salt fog prevention, corrosion prevention, ice protection (if any), and low-temperature protection (if any) than onshore wind turbines. Offshore wind farms are located in different geographical locations, and meteorological parameters such as temperature, pressure, average wind speed, maximum wind speed, maximum wind speed, turbulence intensity, etc., are different. Therefore, when selecting the wind turbine, the range of meteorological parameters in the offshore wind farm area should be clearly defined. This is to ensure the adaptability of the selected wind turbine to the atmospheric environment and to prevent adverse phenomena such as lower availability of wind turbines, higher failure rate of components, and lower life span due to improper selection.

2. **Site and construction conditions**

The water depth is the main factor affecting the offshore wind farm site. Affected by the drafting depth of the lifting and transporting vessel, if the single wind turbine capacity is too large and the parts are too heavy, the draught of the lifting and transporting vessel is too deep to carry out transportation and lifting typically. Therefore, in the areas where the water depth is too shallow, it is not suitable to install wind turbines with excessive single-engine capacity and heavy components. The wind turbine selection is also constrained by the construction conditions, such as lifting and piling capacity. Excessive single-unit capacity will cause difficulties, such as lifting capacity not meeting equipment installation requirements.

3. **Progressiveness and reliability of the technology**

The offshore wind farms are far from the shore, the offshore environment is complex and less accessible, and the operation and maintenance costs are higher. Therefore, the selected wind turbine should ensure a more advanced design concept, more advanced technology, higher equipment reliability, and a certain amount of time of actual operating experience than onshore wind turbines. In this way, it can avoid the decrease in power generation and increase in operation and maintenance costs of offshore wind farms caused by low equipment reliability. Under the premise of guaranteeing the environmental adaptability and safety reliability of wind turbines, meeting the site and construction conditions, and effectively using the wind resources of the proposed offshore wind farm, it is also necessary to choose wind turbines with larger individual capacity as much as possible, so that the offshore wind farm has a higher level of power generation.

4. **Equipment supplier supply and technical service capability**

The construction and operation of offshore wind farms are affected by sea and wind conditions, and the construction has an apparent weather window period. Therefore, the supply capacity of the wind turbine supplier becomes a key factor limiting the construction schedule of offshore wind farm construction, and the selected wind turbine supplier should have sufficient capacity to meet the installation schedule requirements of the wind farm. When selecting the wind turbine, the supplier's technical strength and service level should also be considered. The supplier should be able to cooperate in completing the design and construction of the wind turbine foundation, electrical, and other supporting engineering, and have the ability to guide the lifting and commissioning of wind turbines and rapid response to faults. Therefore, the supplier's supply capacity and technical service level significantly impact the smooth construction of offshore wind farms, optimize the design, shorten the construction period, and reduce the investment and production level of the warranty period.

3.3.5 *Micro Siting Technology for Offshore Wind Farms*

The micro siting technology for offshore wind farms mainly involves the optimal design of offshore wind turbine layouts and the topology optimization of offshore wind farm collector lines. The two aspects constrain and influence each other in terms of cost and generation of revenue. For offshore wind farms, increasing the installation distance between wind turbines while helping to reduce the impact of wake effects will inevitably increase the length of the submarine cable between wind turbines. It raises the investment cost of wind farm collector lines and increases the power loss of collector lines. Therefore, in micro-siting offshore wind farms, joint optimization of offshore wind turbine layout and collector line topology is necessary for the optimal economic balance between tail current impact and submarine cable investment sought.

The joint optimization process of offshore wind turbine layout and collector line topology involves many elements, including the number of wind turbines, wind turbine coordinates, annual wind farm power generation, electrical system costs, operation and maintenance costs, and network losses, which must be weighed comprehensively. To balance the long-term total generation, the benefit of wind farms, and the investment cost of collector lines, a multi-objective planning model for offshore wind farms is usually constructed, which takes into account the wake effect of wind turbines (Generally, the wake simulation is carried out with wake calculation models such as Jensen model, Larsen model, and Frandsen model.), the economics and reliability of collector lines. The best offshore wind turbine layout and collector line topology scheme are available by optimizing wind turbine coordinates and topology.

3.4 Electrical System Design for Offshore Wind Farms

3.4.1 *Offshore Wind Farm Electrical System Composition and Major Equipment*

1. **Electrical system composition of offshore wind farm**

The electrical system of large offshore wind farms mainly includes the power pooling system, offshore substation, and submarine cable for transmission to shore and onshore substation [3]. As shown in Fig. 3.1.

(1) Electricity pooling system

The power pooling system of an offshore wind farm is mainly composed of offshore wind turbine boosting equipment and pooling cables. In an offshore wind farm, each wind turbine needs to be connected to an adjacent wind turbine through a cable

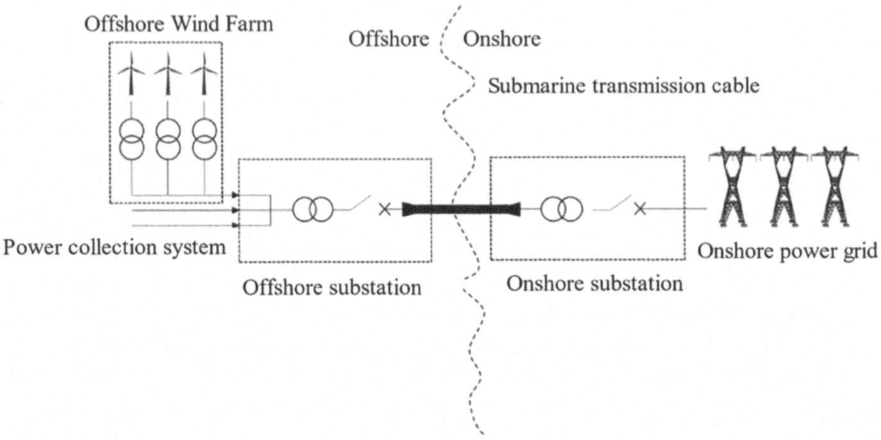

Fig. 3.1 Electrical system for large offshore wind farms

gathered by one or more medium voltage collector switch assemblies and cables. It is finally transmitted to the offshore substation via a voltage boost.

Wind turbines in offshore wind farms are usually divided into several groups, each connected in a star or string pattern. Star-connected wind turbines are first connected to the adjacent transformer-equipped collector platform and then centrally connected to the substation. It has the advantage that each wind turbine does not need an independent transformer, and the cost is low. Still, the construction cost of the collector platform is high, and the system stability is inferior. Each wind turbine in a string connected has an independent transformer, and multiple wind turbines are connected in a string and then pooled to an offshore substation. The two types of connections are shown in Fig. 3.2.

The convergence cables usually converge from 0.69 kV at the wind turbine outlet to 35 kV (typically) within the wind farm. Unlike onshore wind farms, submarine cables connect offshore wind farms to offshore wind turbines. Submarine cables require exceptional performance, such as waterproofing, corrosion resistance, and resistance to mechanical pulling and external collision, but also require high electrical insulation performance and high safety reliability. Based on the complexity of the offshore environment, the gathering cable is generally an optoelectronic composite cable, i.e., a fiber optic unit with a communication function and reinforced structure is added to the submarine power cable to make it have the dual tasks of power transmission and fiber optic information transmission. The submarine photoelectric composite cable can save marine routing resources and reduce manufacturing and offshore construction costs.

(2) Offshore substation

Compared with onshore substations, offshore substations are costly, and the cost of the support structure and installation dramatically exceeds the price of the substation's electrical equipment. From the cost perspective, offshore substations should

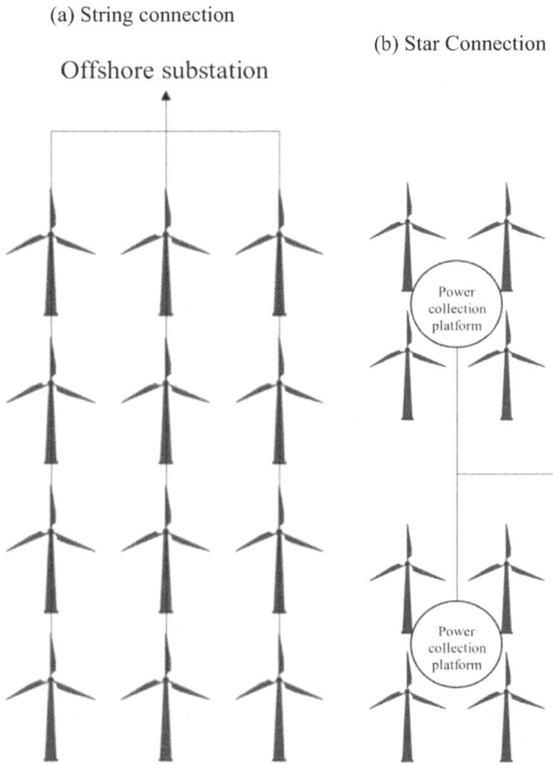

Fig. 3.2 Two types of connections

be avoided as much as possible. For offshore wind farms that are close to land (≤15 km) and have a small capacity (≤100 MW), a 35 kV submarine photoelectric composite cable is usually used to transmit the power directly to the onshore substation. For large-capacity offshore wind farms with long offshore distances, an offshore substation must be set up considering the 35 kV level submarine cable transmission capacity, voltage drop, power factor, and loss. Generally speaking, the power transmission method of offshore wind farms should be planned according to the capacity of the offshore wind farm, the voltage level of the access grid, and the comprehensive cost, which can apply the two-stage rising voltage or three-stage rising voltage method.

(3) Submarine cables transmitted to shore

The voltage level of the submarine cable that transmits to the shore is selected according to different regions, grid types, offshore distance, and installed capacity. Generally speaking, for offshore wind farms with an installed capacity of less than or equal to 100 MW and an offshore distance of less than 10 km, 35 kV collection lines are directly sent to the onshore substation. For offshore wind farms with an installed capacity of 100–500 MW and an offshore distance of 10–100 km, offshore substations are generally installed, and 110 kV or 220 kV high-voltage AC transmission is used to send them to onshore substations. High-voltage DC transmission is the general method for offshore wind farms with an installed capacity greater than 500 MW and an offshore distance of 100 km or more.

(4) Onshore substation

Since it is set up onshore, constructing an onshore substation is more conventional and can be built according to a standardized design. The onshore substation consists of switchgear, measuring devices, transformers, and other equipment. The local grid company can decide on the design. Reactive power compensation devices should also be installed to comply with the grid guidelines and enhance the grid friendliness of the offshore wind farm.

2. **Electrical primary system major equipment**

The leading equipment of the electrical primary system of the offshore wind farm can be divided into four parts: offshore wind turbine, submarine collector line, step-up substation, and plant power system. It is similar to onshore wind farms, mainly including primary equipment such as generators, transformers, submarine cables (current-carrying conductors), reactors, capacitors, current transformers, voltage transformers, high-voltage GIS distribution devices, medium- and low-voltage complete distribution devices, reactive power compensation devices, arc extinguishing, and detuning devices, overvoltage protection and grounding devices.

3. **Electrical secondary system main equipment**

The leading equipment of the electrical secondary system of the offshore wind farm is the low-voltage electrical equipment necessary for monitoring, regulating, controlling, protecting, and providing operating conditions or production command signals for the operation and maintenance personnel. It is similar to onshore wind farms and mainly includes line protection device, transformer protection device, bus difference protection device, submarine cable online monitoring system, fault filter, protection information sub-station, power quality online monitoring device, synchronous vector measurement device, clock alignment system, gateway metering system, GIS equipment local discharge monitoring system, SF6 micro water online monitoring system, leading transformer oil online monitoring system, remote control system, Wind power prediction system, wind farm automatic generation control (AGC) system, automatic voltage control system (AVC), DC system, uninterruptible power supply (UPS) equipment, five-proof system, communication system and monitoring system and other secondary equipment.

The difference between onshore and offshore wind farms is that the electrical secondary system contains an online monitoring system for submarine cables. The submarine cable online monitoring system includes three modules: submarine cable disturbance monitoring system, submarine cable temperature and stress monitoring system, and submarine cable ship monitoring and warning system. It can record the system current, voltage, and derived quantities, such as active and reactive power and system frequency, caused by short-circuit fault, system oscillation, frequency collapse and voltage collapse, and other significant disturbances to reduce the probability of submarine cable failure.

3.4.2 Design of Transmission System for Offshore Wind Farms

1. **Composition of transmission system for offshore wind farms**

Offshore wind farms are generally larger in scale and farther offshore and need to rely on offshore booster substations that use high-voltage submarine cables to send out and connect to the onshore grid. The composition of the offshore wind farm delivery system is roughly the same, mainly containing three parts: offshore booster station, submarine cable, and onshore booster station.

2. **Design of offshore wind farm transmission system**

It needs to meet the technical requirements of active power, power prediction, reactive voltage, fault ride-through, operational adaptability, power quality, simulation model and parameters, secondary system, access grid testing, verification, etc., for offshore wind farm access grid. It must follow the State Grid Corporation corporate standard "Offshore Wind Farm Access System Design Specification" (Q/GDW11411-2015). It needs to meet the design requirements for the offshore wind farm system primary, system secondary, offshore booster station, submarine cable, environmental protection, energy conservation, and emission reduction, as well as coordinate with the power grid development planning in the region.

(1) Offshore wind power access voltage selection

The access voltage level is determined according to the installed size of offshore wind farms and the actual maximum power delivery, combined with the actual transmission distance. When the total scale of offshore wind farms is below 300 MW, they are recommended to be connected to the AC grid of 110 kV and below. When the total scale of the offshore wind farms is 300–600 MW, it is recommended that they be connected to the AC grid of 220 kV and below. When the total scale of offshore wind farms exceeds 600 MW, it is recommended that they be connected to an AC grid of 500 kV and above the voltage level.

(2) Offshore wind farm delivery method selection

Offshore wind farm delivery methods worldwide are divided into high-voltage AC (HVAC) and high-voltage DC (HVDC). HVDC is divided into conventional high-voltage direct current (LCC-HVDC) and flexible high-voltage direct current (VSC-HVDC).

- The HVAC transmission method transmission system structure is simple, and the project cost is low. However, due to the influence of AC cable charging current, a large-capacity dynamic reactive power compensation device, usually used for small-scale and offshore wind farms near the shore, is generally necessary.

- The LCC-HVDC transmission method can be connected to wind farms on a larger scale and farther offshore. It can adapt to various frequency fluctuations in offshore wind farms, not limited by the transmission distance. Transmission losses are low, but the converter station technology is complex and expensive and is generally used for very large offshore wind farms.
- The VSC-HVDV transmission method inherits the advantages of LCC-HVDC. It is based on the solution of LCC-HVDC absorption of large amounts of reactive power, phase change failure, and other problems. It is very suitable for wind farms to connect to the AC grid, but due to the high-power insulated gate type bipolar transistor (IGBT) development level restrictions, the maximum transmission capacity can only reach about 1000 MW at present. It is more suitable for medium and large offshore wind farms.
- The offshore wind farm delivery method design mainly considers factors such as transmission capacity, transmission distance, economy, reliability, and environmental friendliness. The actual project should also be combined with the exact conditions of offshore wind turbines.
- Parameter design of offshore wind farm delivery system [4].
- The selection of cable cross-sectional area and the number of outgoing circuits. The selection principle of cable cross-sectional area: the long-term permissible current of the cable is the basis for selecting the cross-sectional area of the cable. Suppose the long-term permissible current of the cable is equal to the long-term operating current of the cable. In that case, the selected cable cross-sectional area meets the requirements of the system transmission capacity. The selection principle of the number of outgoing circuits: the transmission capacity of the transmission line needs to meet the wind farms' entire generation needs, combined with the regional grid operating requirements, to clarify whether the transmission line needs to meet the "N – 1" principle.
- Offshore booster transformer requirements. The selection of offshore booster transformers mainly involves capacity, type, and wiring type. The selection of the offshore step-up transformer capacity needs to refer to the adequate capacity of the offshore wind farm and take into account the offshore wind farm's planned capacity and phasing scale. For AC step-up transformers, 220 kV can be considered but not limited to 180 and 240 MVA, and 500 kV can be considered but not limited to 750 and 1000 MVA. For DC converter transformers, the total transformer capacitance should be at least the wind farm's adequate capacity. The offshore step-up transformer should be an on-load voltage regulating transformer. The primary connection type can consider unit and single bus connections and can be flexibly selected based on needs.
- Reactive power compensation device. When the reactive power capacity of offshore wind turbines cannot meet the demand of system voltage regulation, appropriate capacity reactive power compensation devices should be installed centrally in offshore wind farms, and dynamic reactive power compensation devices should be installed when necessary. Reactive power compensation devices

can be considered for installation in offshore and onshore substations. According to the principle of power system reactive power compensation by voltage layer and power supply area, a specific capacity of inductive compensation device needs to be considered to be configured to absorb the charging power of submarine cables. At the same time, a particular capacity of dynamic reactive power compensation should be configured to facilitate wind farm side voltage regulation and power factor adjustment.

- High impedance configuration scheme for transmission lines. For the AC transmission scheme, the charging power generated by the AC submarine cable should be compensated by installing shunt reactors to ensure the reactive power balance of the submarine cable and minimize the reactive power transmitted through the line.
- Requirements for switchgear. Switching equipment should meet the corresponding access to the grid voltage level switch short-circuit current requirements; the specific reference is as follows: The short-circuit current level for 35 kV access to the grid can be selected as 31.5 kA or 40 kA. The short-circuit current level for 220 kV access to the grid can be chosen as 40 kA or 50 kA. The short-circuit current level for 500 kV access to the grid can be selected as 50 kA or 63 kA.

3. **Offshore Boost Station Design**

According to the difference in transmission method of offshore wind farm transmission systems, offshore booster stations are mainly divided into substations based on the HVAC transmission method and converter stations based on the HVDC transmission method. The substations and converter stations on the transmission side are mainly built at sea or on offshore islands and connected to offshore wind turbines. The substations and converter stations on the receiving side are constructed primarily on land and connected to the grid. Since the requirements for onshore substations and converter stations are the same as those for conventional onshore wind farm construction, the relevant construction specifications are more complete. Therefore, the design of offshore substations and converter stations is mainly elaborated.

(1) Offshore substation design

Offshore substations aggregate the power of offshore wind turbines. After boosting, it is transported through submarine cables to the onshore centralized control center and integrated into the power grid, which mainly improves power transmission efficiency. The design of an offshore substation mainly involves an electrical main wiring scheme, selection of leading electrical equipment, insulation fit and overvoltage protection, secondary system scheme, lightning protection, grounding design, offshore substation structure design, etc.

1) *Electrical main wiring scheme*

The selection of the electrical main wiring scheme for offshore substations requires consideration of the total installed capacity of the offshore wind farm, the number of main transformers, the voltage level, and the number of outlet circuits. It mainly

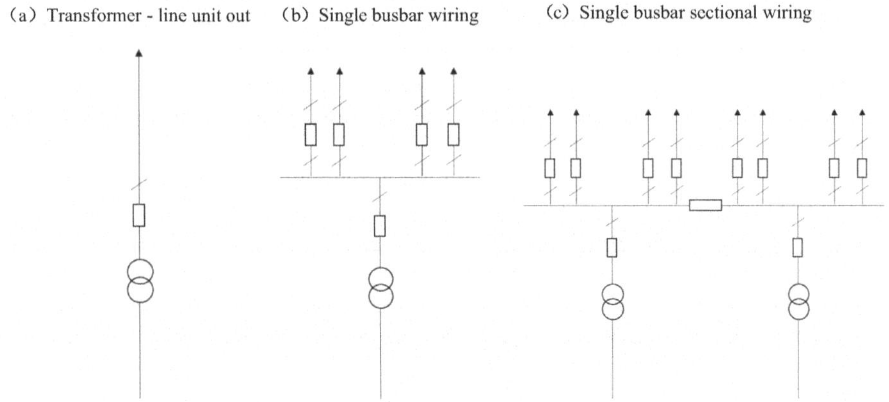

Fig. 3.3 Three kinds of wiring scheme

includes transformer-line unit out, single busbar wiring or single busbar segmentation wiring, and three other forms. The three wiring schemes are shown in Fig. 3.3.

The advantages of the transformer-line unit outlet are simple wiring, minimal equipment, no need for high-voltage distribution equipment, and savings in platform space. The disadvantage is insufficient electrical reliability. It is suitable for single or two main transformer-scale projects.

The advantages of single busbar wiring are simple wiring, clear wiring, less equipment, easy operation, saving investment, and easy expansion and use of complete sets of distribution equipment. The disadvantage is that it is not flexible and reliable enough and requires high-voltage power distribution devices. It is suitable for projects of a single main transformer scale.

The advantage of single bus section wiring is that when a bus fault occurs, the section circuit breaker automatically removes the fault section to ensure the regular bus section has an uninterrupted power supply. The disadvantage is that when a bus or bus disconnect switch fails, or maintenance, the circuit of the section of the bus needs to be shut down during the maintenance period. When the outgoing line is a double circuit, it often makes the overhead line cross over, and the expansion needs to be balanced in both directions. It is suitable for single or two main transformer scale projects.

2) *Selection of main electrical equipment*

The offshore substation platform is located in an offshore environment. Therefore, in addition to meeting national and general power industry standards, the electrical equipment in the station also needs to meet the unique requirements for equipment in humid and highly corrosive environments. The selection of leading electrical equipment for offshore substations mainly involves main transformers, high-voltage gas-insulated (GIS) power distribution devices, reactive power compensation devices, medium and low-voltage power distribution devices, rotating electrical equipment, and lighting.

- Main transformer: According to the installed capacity of offshore wind farms and environmental conditions, the main transformer should be selected as a three-phase, on-load regulator transformer using the natural wind cooling method. Large-capacity transformers should use forced oil circulation wind cooling. The sending line should be an enclosed installation with no externally exposed charged part, the joint part of the soft connection method. It must be equipped with a space heating device; the shell is made of corrosion-resistant materials to prevent moisture and condensation of equipment.
- High-voltage gas-insulated (GIS) power distribution units: They have the advantages of compact structure, small footprint, high reliability, flexible configuration, easy installation, strong environmental adaptability, low maintenance workload, etc. It is especially suitable for the marine environment, which has high humidity and heavy salt spray and is affected by typhoons. The rated voltage, rated current, and rated opening current need to be configured and selected according to the voltage level of the main wiring scheme, the design transmission capacity of each circuit, short-circuit current, etc.
- Reactive power compensation devices: Offshore substation equipment is limited by the marine environment and installation space, so it generally chooses a relatively small size of SVG dynamic reactive power compensation device. The reactive power source of offshore wind farms is mainly high-voltage submarine cables and collector lines, and the main equipment for reactive power consumption is the transformer. The adjustment range of the reactive power compensation device should be selected according to the amount of reactive power compensation when the transformer is running at no load and full load.
- Medium/low voltage power distribution devices: The rated voltage, rated current, and rated breaking current need to be configured and selected according to the voltage level of the main wiring scheme, the designed transmission capacity of each circuit, and the short circuit current. They must be placed indoors and equipped with a space heating device. The enclosure is made of corrosion-resistant materials to prevent moisture and condensation on the equipment.
- Rotating motor equipment, including diesel generators (emergency power, station power), electric pumps, etc. The rated power and rated voltage are selected according to the required power and voltage level of each rotating motor equipment, which must be equipped with space heating devices to prevent moisture and condensation. When used on offshore platforms, fully enclosed motors equipped with all cast metal parts, non-corrosive and non-sparking cooling fans, anti-corrosion hardware, and stainless steel nameplates are usually used.
- Lighting: mainly divided into standard working lighting and accident lighting. Regular working lighting power supply by the station AC power supply. According to the general lighting settings, individual places set local lighting. The power supply of accident lighting is supplied by the AC and DC switching device in the station distribution panel, and a certain number of emergency lights and indicators need to be set up in the stairs and corridors. Lighting fixtures, switches, and sockets need to be made of moisture-proof, corrosion-resistant, and high humidity-resistant materials, and it is recommended to add mechanical protection.

3) *Insulation coordination and overvoltage protection*

The insulation fit and overvoltage protection of offshore substations should fully consider the characteristics of long submarine cable feeders for offshore wind farms. Comply with the insulation fit principles stipulated in the IEC 60071 "Insulation Fit" series of standards and GB/T 50064–2014 "Design Code for Overvoltage Protection and Insulation Fit of AC Electrical Installations" and other domestic and international codes for design. Configure appropriate overvoltage protection devices, and the insulation fit margin between the overvoltage level, equipment insulation level, and protection device characteristic parameters should meet the specification requirements [4].

4) *Secondary system scheme*

Offshore substations are characterized by unattended, long offshore distances, poor operating environment, inconvenience in maintenance, and significant failure losses. The secondary system equipment is similar to traditional onshore substations. Still, it has higher standards and requirements, such as a more compact equipment layout, high requirements for equipment salt fog resistance, high requirements for equipment moisture resistance, more complete monitoring system functions, higher reliability of telecontrol and communication equipment, higher requirements for automatic fire alarm, longer backup time for DC systems and uninterruptible power supply (UPS) systems, and higher reliability requirements.

5) *Lightning protection and grounding design*

The offshore substation shall be designed for lightning protection by the standard IEC 62305 "Lightning Protection." It mainly includes direct lightning protection design, lightning surge overvoltage protection, inductive lightning protection design, and offshore substation grounding design.

6) *Structural Design of Offshore Substation*

The structure of offshore substation mainly includes two parts: upper platform structure and lower foundation structure. The leading electrical equipment is arranged on the upper platform, mostly a steel structure, to reduce weight. The lower foundation structure needs to be selected according to the specific submarine foundation, seawater depth, platform weight, layout, and transportation and installation requirements, mainly divided into the single-pile foundation, multi-pile foundation, conduit frame foundation, etc. Both the upper platform structure and the lower foundation structure need to be designed for corrosion prevention, usually using coating and cathodic protection anti-corrosion technology; the foundation structure coating anti-corrosion design is generally considered according to 27 years and reserved for the atmospheric zone, wave splash zone, underwater zone, and mud under the zone and another single-sided corrosion margin.

(2) Offshore converter station design

HVDC transmission-based offshore converter stations are divided into LCC-HVDC transmission-based offshore converter stations and VSC-HVDC transmission-based offshore converter stations. Compared with the LCC-HVDC transmission method, the VSC-HVDC transmission method has the advantages of flexible control, small footprint, modular design, low noise, etc. It has more competitive advantages in offshore wind power transmission. Therefore, this section focuses on the design of the offshore converter station based on the VSC-HVDC transmission method. The primary transformer selection, AC equipment selection, rotating motor, lighting, lightning protection, grounding, and structural design of the offshore converter station are the same as the design principles of the offshore substation and will not be repeated in this section.

The design of the offshore converter station based on VSC-HVDC transmission mode mainly involves the topology of the primary circuit of the flexible DC converter, the main wiring scheme, the plane setting of the converter station, the selection of the leading electrical equipment, overvoltage and insulation, and the secondary system scheme.

1) *The topological structure of the primary circuit of the flexible DC converter*

The flexible DC converter's primary circuit mainly includes three topologies: a two-level voltage source converter, a three-level voltage source converter, and a modular multi-level voltage source converter, as shown in Fig. 3.4.

Currently, most practical projects of flexible DC transmission adopt two-level and three-level voltage source converter NPC structures. Although it has many advantages over the traditional thyristor high-voltage direct current transmission, it also has the disadvantages of significant loss in the converter station and failure to control the fault current when the DC side is faulty. The domestic promotion has specific difficulties [only ABB company has a mature series insulated gate bipolar transistor (IGBT) dynamic voltage equalization technology], and there is a need to configure AC filters and other defects.

Modular multilevel voltage source converter topology is the most feasible for domestic flexible DC transmission. Modular multilevel voltage source converter topology has low loss and modularization, allows the use of more mature standard components, has strong fault protection capability, can relieve the electrical pressure on the converter valve, and does not require filters, etc. The disadvantage of the modular multilevel voltage source converter topology is that the sinusoidal bridge arm current is distorted due to the existence of the internal circulation of the converter, the rated current requirements of the switching devices are increased, and the controller is more efficient than the two-level and three-level NPC the structure is much more complicated.

2) *Main wiring scheme*

The HVDC transmission system comprises a rectifier station (sending end), an inverter station (receiving end), and a DC line. The characteristics of the VSC-HVDC transmission mode determine that it is easy to realize multi-terminal direct current transmission, and its converter station can operate as both a rectification station and

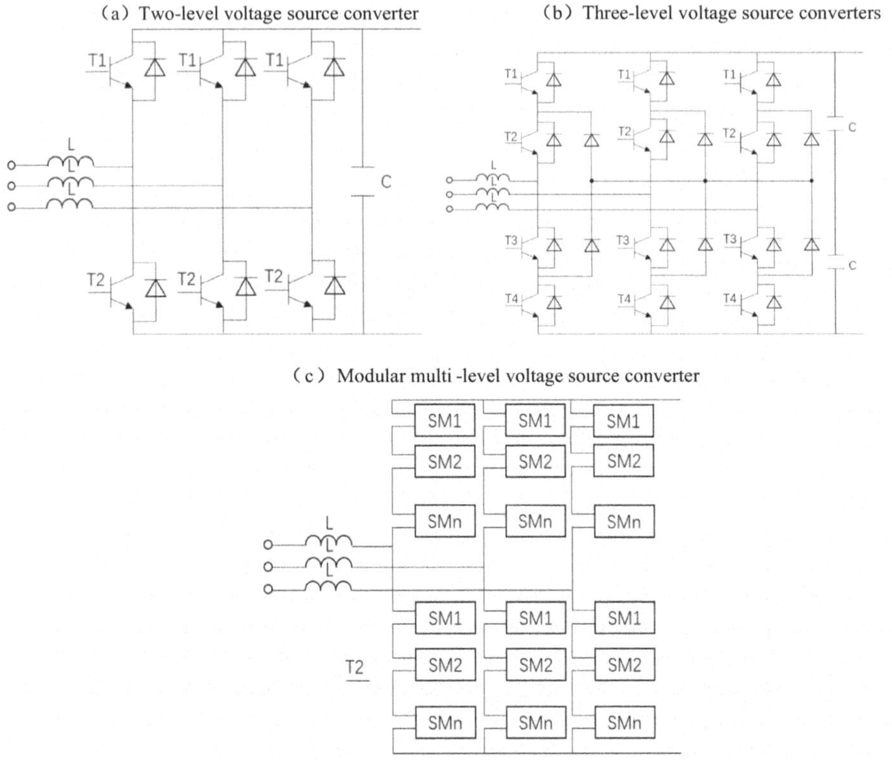

Fig. 3.4 Flexible DC converter main circuit topology

an inverter station. The rectifier station that operates during forward power transmission becomes an inverter station during reverse power transmission. In contrast, the inverter station that operates during forward power transmission becomes a rectifier station during reverse power transmission. The main wiring and primary equipment of the rectifier station and the inverter station are the same.

The main wiring scheme of the VSC-HVDC transmission system mainly involves the AC side wiring mode and the DC side wiring mode. The AC side wiring method primarily includes the connection method for connecting the transformer, reactor, and AC filter. The DC side wiring methods mainly include the monopolar ground circuit operation mode, the single-pole metal return line, the neutral point grounding at both ends of the bipolar pole, the neutral point grounding at one end of the bipolar pole, the IEC bipolar wiring mode, and two converter units series, etc.

3) *The plane setting of the converter station*

The layout of the offshore converter station should be considered from the perspective of compactness based on the electrical main wiring scheme, construction conditions, noise control, and the direction of outgoing lines. According to the functional orientation and working principle, the converter station can be divided into four major areas: AC field, valve hall, DC field, and functional room.

The AC field area mainly includes AC incoming lines, transformers, starting circuits, and reactors. The valve hall area mainly includes converters, valve hall grounding knife gates, etc. The DC field area mainly includes DC reactors, isolating switches, current measuring devices, voltage measuring devices, lightning arresters, DC PLC (reserved) fault location devices, etc. The functional room area mainly includes a low-voltage power distribution room, relay room, central control room, valve cooling equipment room, spare parts room, communication room, secondary battery room, communication power room, tool room, etc. In addition to laying out the technological process design according to the content of the four major areas, the offshore converter station should consider the multi-layer layout instead of the single-layer layout commonly used in onshore converter stations to save floor space.

4) *Selection of main electrical equipment*

The selection of the offshore converter station's leading electrical equipment mainly involves selecting voltage source converter valve development components, DC capacitors, and parameters for connecting transformers and phase reactors, as well as other leading equipment.

- Selection of development devices for voltage source converter valves: The transistor power device IGBT has the advantages of stable and reliable operation and a wide range of suppliers, which can meet the application of most high-voltage and large-capacity converters, and occupies a dominant position in VSC-HVDC power transmission projects. It mainly includes pressure-connected IGBTs suitable for two-level, three-level topologies and cascaded two-level topologies, modular multi-level voltage source converter topologies, and modular IGBTs for similar topologies. When selecting the IGBT type, it should be checked that the voltage and current that the IGBT withstands under each working condition should not exceed the safe working area, and a certain safety margin should be left.
- Selection of DC capacitors: DC capacitors mainly include those for switching VSC converter stations and modular multi-level current source converters. The main design principles are to ensure that energy storage can support DC voltage and have low stray inductance. At the same time, the fluctuation of DC voltage should be suppressed within the allowable range, weakening the coupling effect between two converter stations and matching the dynamic response speed with the control system.
- Selection of connection transformer and phase reactor parameters: Transformers and phase reactors are connected to jointly limit the threshold arm or the fault current in the event of a DC bus short circuit. In engineering design, the transformer's short-circuit impedance is generally fixed first. Then, the phase reactor's parameters are determined by considering various conditions, such as "DC short-circuit current limitation."

- Selection of other leading equipment of the converter station: In addition to the above equipment, VSC-HVDC power transmission should also include switchgear, neutral point grounding branch, filter, DC reactor, measuring device, and other equipment.

5) *Overvoltage and insulation coordination*

The configuration principle of the lightning arrester in the flexible DC converter station is the same as that of the ordinary HVDC converter station. The AC side arrester should limit the overvoltage generated on the AC side, and the DC side arrester should limit the overvoltage generated on the DC side. The overvoltage check of the system should select and analyze the overvoltage situation after the arrester is installed in the most severe working condition. When choosing the insulation coordination margin, the AC side shall be designed according to the standard design configuration of the corresponding voltage level AC substation, and the DC side shall be designed according to the level of the DC converter station with an insulation margin more significant than 500 kV.

6) *Secondary System Solution*

The secondary system scheme of the offshore wind farm converter station differs from that of the traditional onshore wind farm converter station. It is the same as the secondary system scheme of the offshore wind farm substation.

4. **Submarine cable laying design**

Submarine cable laying design mainly includes submarine cable route selection, submarine cable type selection, submarine cable grounding method selection, submarine cable laying, protection, etc.

(1) Submarine Cable Routing Selection [4]

1) The landing point of the submarine cable should be selected on the coast where the wind and waves are relatively stable throughout the year and the sea and tide are relatively weak. Choose a coast suitable for the submarine cable to land vertically as soon as possible to reduce the length of laying parallel to the coastline and stay away from areas prone to natural disasters. Choose a stable coast suitable for construction to avoid existing and planned development zones, chemical plant areas and heavily polluted areas, nature reserves, scenic spots, etc., and avoid built or planned power cables, communication submarine cables, petroleum pipelines, gas pipelines, water supply and drainage pipes and other obstacles.
2) Submarine cables should be routed with flat underwater terrain and avoid areas where the seabed is bedrock. In addition, you should also prevent natural obstacles on the seabed, reef areas, areas where the seabed topography fluctuates sharply, regions where the seabed moves and scours violently, sandy slope areas, and areas with tidal solid and sandy undercurrents.

3) Submarine cables should fully consider the impact of various construction projects existing and planned by other relevant departments:
   ① Active development areas on the sea should be avoided, such as port development areas, planned construction areas, reclamation areas, offshore oil platforms, etc.;
   ② Strong exclusive marine functional areas should be avoided, such as naval training bases or test bases or testing areas, dredging sites, garbage dumping sites, etc.;
   ③ Obstacles such as sunken ships and underwater structures should be avoided;
   ④ Areas of fishery activities such as aquaculture and professional fishing should be avoided, and if it is impossible to avoid, necessary protective measures should be taken;
   ⑤ Avoid waters where ships often drop anchor, stay away from anchorages and busy waterways, set up no-anchor areas for unavoidable waterways, and take necessary submarine cable protection measures and early warning measures;
   ⑥ It should be as far away from other submarine pipelines as possible. The horizontal distance should not be less than the following values: 500 m in coastal broad sea areas, 100 m in narrow sea areas such as bays, and 50 m in harbor areas; try to avoid crossing with other pipelines if it is unavoidable, you should take necessary security measures.

(2) Submarine cable selection

Submarine cable selection mainly includes insulation type, conductor cross-sectional area, metal sheath, outer sheath, armor type, etc.

1) AC cables can be divided into oil-filled cables and Cross-linked polyethylene insulated cable (XLPE) insulated cables according to insulation types; DC single-core cables can be divided into oil-filled cables, XLPE insulated cables, and non-drip paper insulated cables according to insulation types wait.
2) The selection principle of the conductor's cross-sectional area is that the cable's long-term allowable current should meet the continuous working current requirements, and the short-circuit thermal stability requirements should be met during short-circuit. A voltage drop check should be performed if necessary, depending on the cable length.
3) The metal sheath of submarine cables is generally made of the lead sheath, which has good sealing performance, can prevent moisture or moisture from entering the cable insulation, and has good corrosion resistance.
4) Outside the metal sheath, an outer sheath needs to be extruded, and its primary functions are compression resistance, water resistance, moisture resistance, and mechanical protection. For short-distance submarine cables, if the shock-induced overvoltage in the metal sheath does not pose a threat to the outer sheath, polyethylene (PE) outer sheath can be used; when the submarine cable is long, the shock-induced overvoltage in the metal sheath When the voltage is

too large, it is advisable to use semi-conductive polyethylene with added carbon black as the outer sheath.

5) Armoring is a crucial structural component of submarine cables, maintaining tension stability and providing mechanical protection. The design of the armor layer should meet the requirements for the mechanical tensile strength of the submarine cable under the conditions of laying, operation, maintenance, and salvage. AC cables have circulation loss, hysteresis loss, and eddy current loss in the armor, and the materials used need to be considered comprehensively. In DC cables, the armor material only needs to meet the requirements of mechanical properties and anti-corrosion. Commonly used submarine cable armor mainly includes galvanized thick round steel wire armor, galvanized flat steel wire armor, stainless steel wire armor, and flat copper wire armor.

(3) Selection of Grounding Mode for Submarine Cables

Submarine cable lines usually adopt three grounding methods: direct grounding at both ends without short-circuiting in the middle, direct grounding at both ends, and segmental short-circuiting and direct grounding at both ends with a semi-conductive outer sheath [5].

① For AC submarine cable lines, under the premise of meeting the limit value of power frequency induced voltage. Suppose the impulse induction voltage of the metal sheath caused by the intrusion wave of overvoltage exceeds the impulse withstand voltage of the insulating outer sheath. In that case, the impulse induction voltage must be limited. The grounding measures of "directly grounding at both ends, segmented short-circuiting" or "directly grounding at both ends, using a semi-conductive outer sheath" should be adopted. Suppose the impulse induction voltage of the metal sheath caused by the intrusion wave of overvoltage does not exceed the impulse withstand voltage of the insulating outer sheath. In that case, the grounding method of "directly grounding at both ends and not shorting in the middle" can be adopted.

② For DC submarine cable lines, there is no induced voltage in the metal sheath under normal circumstances, and the grounding method can be directly grounded at both ends. Under the action of lightning or operation impulse overvoltage, a higher induced overvoltage will appear in the metal sheath, and the longer the line, the higher the overvoltage amplitude. When the line reaches a certain length, the cable's outer sheath may be broken down due to excessive shock-induced overvoltage. It is necessary to adopt the grounding measures of "directly grounding at both ends, segmented short circuit" or "directly grounding at both ends, using a semi-conductive outer sheath."

(4) Submarine cable laying and protection

Submarine cable laying and protection measures mainly include buried protection, trench protection, wearing pipe protection, and covering protection [4].

① Buried protection: Use professionally designed cable burying machinery to bury submarine cables to a depth below the surface of the seabed where standard fishing gear and anchors cannot reach to protect submarine cables from external risks to the greatest extent.

② Trench protection: When the water depth of the landing section and the shoal area is shallow, the laying ship cannot approach, and the buried machinery is challenging to construct; the cable trench can be dug before the cable laying ship lays the cable.

③ Wearing pipe protection: Submarine cable routed offshore in a shallow section frequented by fishing activities is a frequent point of anchoring fishing vessel operations. When the submarine cable burial depth does not meet the requirements, iron sheath protection and pre-buried copper or reinforced concrete pipe can be used for protection.

④ Covering protection: When the seabed is a reef, and it is difficult to bury, the burial depth does not meet the design requirements, or it cannot be buried because the submarine cable intersects with other pipelines, the cover protection method can be adopted, that is, the submarine cable is covered with rocks, concrete blocks, sandbags, etc. Cover it to play a protective role. Standard methods include riprap protection, concrete pad protection, and concrete bag (sandbag) protection.

3.5 Foundation Structure Design of Offshore Wind Turbine and Booster Station

3.5.1 Classification and Design of the Foundation Structure of Offshore Wind Turbine and Booster Station

As a supporting structure, offshore wind turbines play a vital role in the safe operation of wind power systems. In the design process, it is necessary to fully consider the influence of external environments such as offshore distance, seabed geological conditions, offshore wind waves, currents, and sea ice. Therefore, the basic structure of offshore wind turbines is more complex and diverse than onshore wind turbines. According to the different ways of fixing the seabed, it can be divided into two categories: fixed and floating. The fixed type is generally used in shallow seas, suitable for water depths of 0–50 m, and the structural types are mainly divided into pile-foundation and gravity-type. The floating type belongs to the deep water structure type and is used primarily in waters with a more than 50 m depth. Compared with the foundation structure of offshore wind turbines, the foundation of the offshore booster station bears a larger volume and weight. It has higher requirements for the foundation structure's volume, strength, and stiffness. Multi-pile foundations and conduits in pile-supported foundations are usually used as frame foundations. Affected by differences in sea environment conditions, multi-pile foundations are used primarily in intertidal sea areas; jacket foundations are mainly used in offshore sea areas with complex seabed conditions.

1. **Piled foundation**

The pile foundation has the advantages of high bearing capacity, good stability, a small settlement, a light structure, and less resistance to waves and currents. It is suitable for various geological conditions that can sink piles, especially for soft soil foundations. On the rock foundation, pile foundations are used when the thickness of the covering layer is appropriate, and rock-socketed piles can be used when the covering layer is thin. The disadvantage of the pile-supported foundation is that it is subject to great constraints of seabed geological conditions and water depth, and is prone to bending deformation. It requires special piling equipment to be installed, the construction and installation costs are high, and it is sensitive to erosion. Pile-supported foundations can be divided into single pile foundations, tripod foundations, jacket foundations, and group pile-bearing foundations according to the number of pile foundations and connection methods [6].

(1) Single pile foundation

The single-pile foundation is composed of welded steel pipes, and the pile foundation and the tower are connected by welded steel pipes or casing flanges, which is the simplest structure type in the pile-supported foundation. The monopile foundation has a simple structure, saves materials, and has a simple manufacturing process. It does not need any seabed preparation. The construction process is relatively simple and suitable for medium water depths below 30 m. Therefore, it is sensitive to scouring, vibration, and verticality and requires special installation equipment. It is not suitable for locations with many large boulders in the seabed and is challenging to move. The single pile foundation structure is shown in Fig. 3.5.

(2) Tripod foundation

The tripod foundation is also called the multi-pile foundation. It comprises a central column, three copper pipe piles inserted into the seabed at a certain depth, and a strut. It can overcome the disadvantages of oversized pile diameter of monopile foundations and the need for scouring protection. It has better stability and is suitable for areas with water depths above 30 m. Because the same geological conditions of the seabed are more constrained, it is ideal for the sea bed is more brutal, unsuitable for shallow sea, soft seabed, high construction cost, and difficult to move. The tripod foundation structure is shown in Fig. 3.6.

(3) Jacket foundation

The jacket foundation is the most commonly used foundation structure for offshore platforms. It has been maturely applied in the construction of deep-sea oil production platforms. It is mainly composed of jacket and steel pipe piles. The jacket is a steel cone-shaped space frame with steel pipes as ribs. It is a prefabricated steel component and can be designed into a three-leg, four-leg, three-leg plus center pile, four-leg plus center pile, etc., generally composed of cylindrical copper pipes. The jacket and steel pipe piles are typically connected at the surface of the seabed and driven into the seabed through the steel pipes at each leg of the jacket. The jacket foundation can effectively improve the rigidity of the supporting structure and is suitable for large-capacity offshore wind turbines and offshore booster stations. It has low requirements

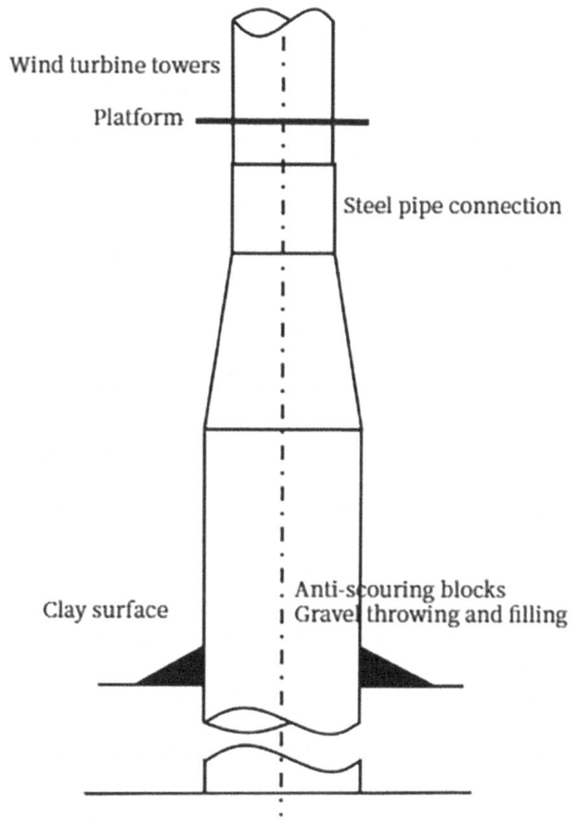

Fig. 3.5 Single pile foundation structure

for geological conditions and is less affected by waves and currents. The technology is mature, and the cost increases exponentially with increased water depth. The base structure of the conduit frame is shown in Fig. 3.7.

(4) Group pile-bearing foundation

The group pile-bearing foundation mainly comprises steel pipe piles and platform supports. The platform support adopts a reinforced concrete cast-in-place structure. Group pile foundation has the advantages of high structural rigidity, controlled construction risk, low total cost, resistance to ship impact, and relatively low requirements for piling accuracy. It has more construction procedures, more self-importance, more piles required, and a large workload of casting in place, and it is mainly suitable for offshore wind farms with a water depth of 0 ~ 25 m and not far from the offshore distance. The pile cap foundation structure is shown in Fig. 3.8.

2. **Gravity foundation**

Gravity foundation is a traditional foundation type, generally reinforced concrete structure, which is the foundation with the most significant volume and mass among

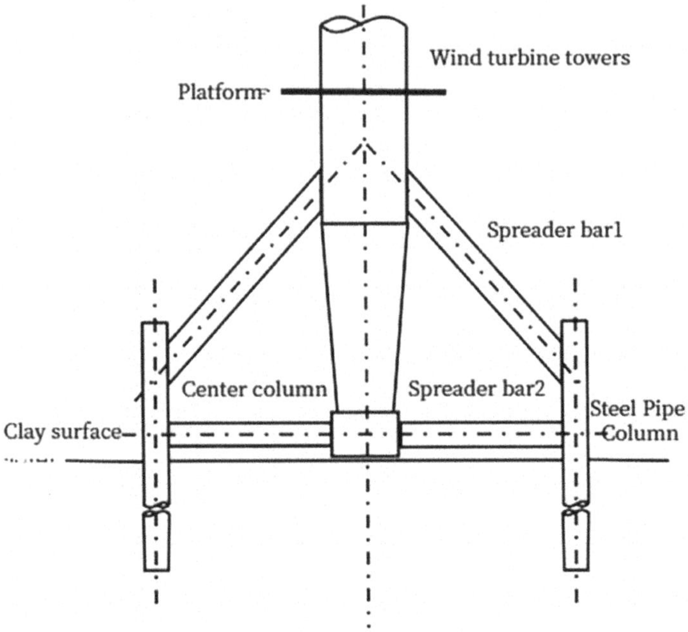

Fig. 3.6 Tripod foundation structure

all foundation types. It relies on its gravity to resist the overturning moment and the load transmitted to the foundation by the superstructure, thus maintaining the stability of the overall structure of the offshore wind turbine and the booster station. Gravity foundation has a simple structure, low cost, good resistance to storms and wind and wave attacks, good stability and reliability, requires ballast, and mature installation technology. However, the seabed needs to be cleaned up during construction, which is greatly affected by wave erosion, volume, and quality. It is significant, difficult to remove, and only suitable for shallow waters. Gravity foundations can be divided into caisson, large-diameter cylinders, and suction foundations according to different wall structures [1].

(1) Caisson foundation

A caisson foundation is a giant empty box of reinforced concrete or steel. The box inside is divided into several cabins with vertical and horizontal partition walls. Generally, after being prefabricated by a particular prefabrication factory, airbags or slides are used to slide down the water with a trolley. When there are few prefabricated caissons, the local ship repair yard, slipway, slipway, airbag, or other suitable natural shore beaches can be prefabricated and launched into the water. After positioning in the water, it is sunk on the leveled bed using the flood ballast method and then filled with sand or block stone inside the caisson. The caisson foundation has the advantages of a small underwater workload, good structural integrity, seismic solid performance,

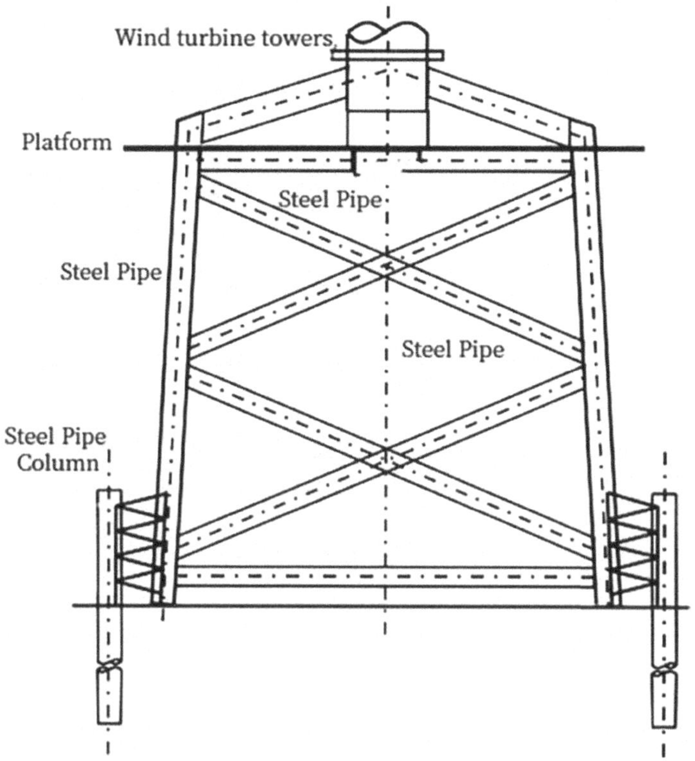

Fig. 3.7 Jacket foundation structure

and fast construction speed. However, the steel consumption is significant, and unique construction equipment and suitable construction conditions are required.

(2) Large-diameter cylinder foundation

The wall body of the large-diameter cylinder foundation is a prefabricated large-diameter, thin-walled reinforced concrete bottomless cylinder. The cylinder is filled with stones, sand, or soil, mainly relying on the gravity formed by the cylinder and part of the filling to resist the load acting on the foundation. The cylinder can be sunk directly into the foundation or rested on a riprap bed. Compared with the caisson foundation, the large-diameter cylinder foundation has a simple structure, less concrete and steel, and strong adaptability to the foundation. It is unnecessary to make a riprap foundation bed, which is cheap and fast in construction. The base pressure generated by the cylinder is enormous, and the construction of the large-diameter cylinder foundation that needs to be sunk into the foundation is more complicated. The basic structure of the large-diameter cylinder is shown in Fig. 3.9.

(3) Suction foundation

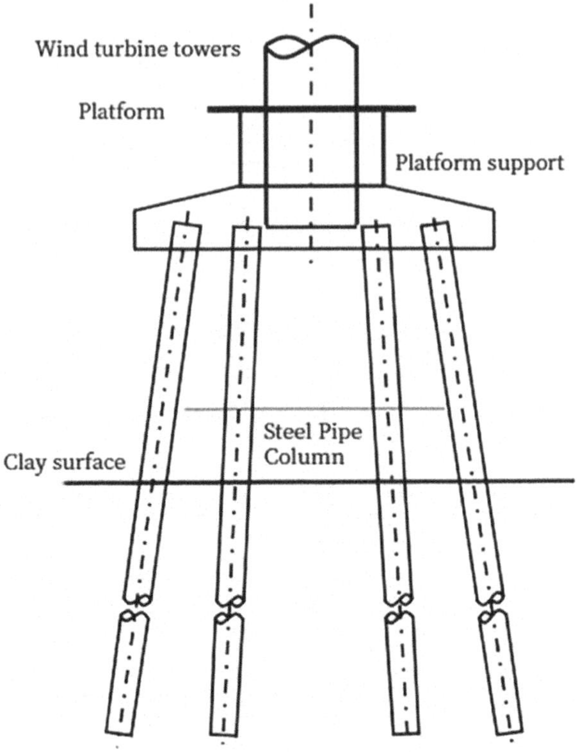

Fig. 3.8 Group pile-bearing foundation structure

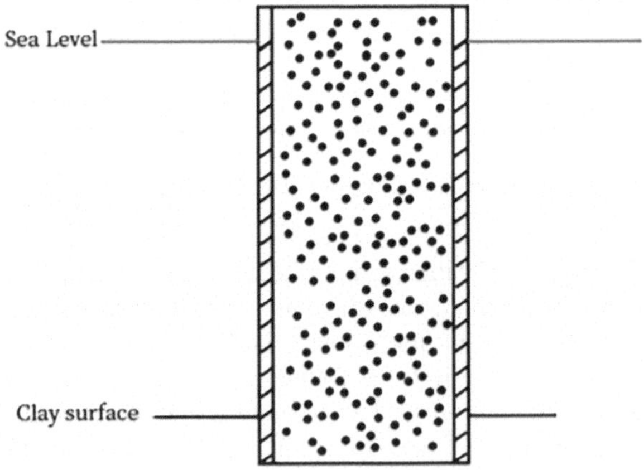

Fig. 3.9 Large diameter cylinder base structure

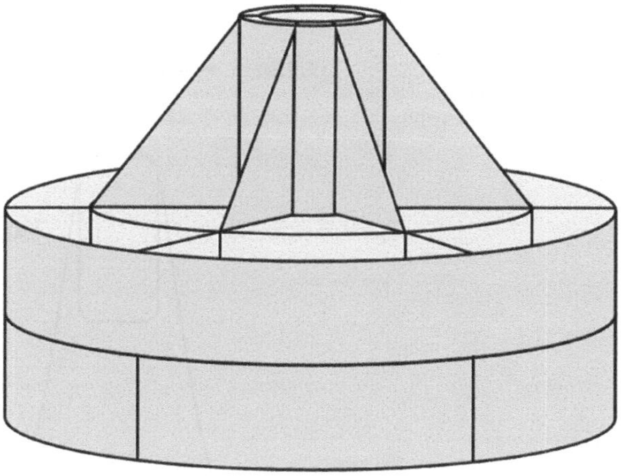

Fig. 3.10 Prestressed concrete structure suction foundation structure

The suction foundation is a large-diameter drum structure with an open and closed bottom. The top of the drum is equipped with a water outlet hole that can be connected to the pump system, and it is usually a steel structure or a reinforced concrete structure. The suction foundation transmits the vertical and horizontal force to the foundation and surrounding soil through the large contact surface between the top cover, the side skirt, and the soil. The suction foundation has the characteristics of simple construction, flexible maneuvering, safety and reliability, and recyclable reuse. It is suitable for deep and shallow seas, and the geological conditions are preferably sandy soil or soft clay. The suction foundation structure of the prestressed concrete structure is shown in Fig. 3.10.

3. **Floating foundation**

Offshore wind power from the intertidal and offshore to the deep sea is the inevitable development trend of the offshore wind power industry. The manufacturing cost and installation difficulty of fixed foundations such as pile-supported and gravity-type foundations are getting higher and higher, which cannot meet deep-sea wind farms' construction and economic requirements. The floating foundation uses an anchoring system to anchor the floating body structure to the seabed. As a primary platform for installing offshore wind turbines, it is especially suitable for sea areas with a more than 50 m water depth. It has the characteristics of low cost and convenient transportation. The floating foundation is still in the research stage and has no large-scale commercial application. It can be divided into Spar-buoy, tension leg platform, and semi-submersible platform.

(1) Spar-buoy

The Spar-buoy lowers the center of gravity of the entire system below the center of buoyancy through ballast tanks to ensure the stability of the whole wind turbine in water and then maintains the position of the wind turbine through radially arranged catenary lines. The Spar-buoy has a large draft, a small vertical wave excitation force,

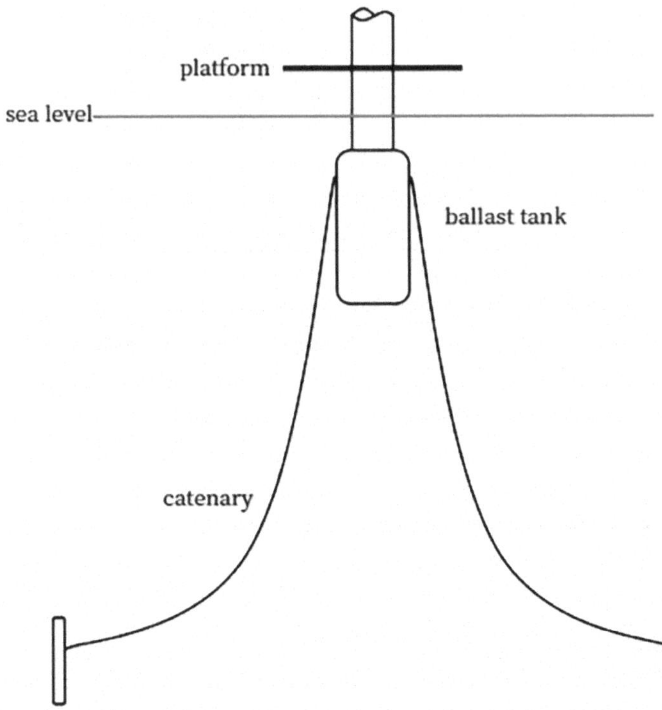

Fig. 3.11 Spar-buoy

and a small heave motion. It has better heave performance, but the contribution of the waterline to stability is small, and the roll and pitch values are significant. Figure 3.11 shows the Spar-buoy.

(2) Tension leg platform

The tension leg platform mainly comprises a cylindrical central column, a pontoon with a rectangular or triangular section, and an anchoring system. The tension leg foundation has good heave and rocking motion characteristics. Still, it has the disadvantages of a complex tension mooring system, high installation cost, considerable tension affected by ocean currents, and resonant motion due to frequency matching of superstructure and mooring system. The tension leg foundation structure is shown in Fig. 3.12.

(3) Semi-submersible platform

The semi-submersible platform ensures the stability of the entire wind turbine in the water through the buoyancy tank located on the sea surface. Then, it maintains the wind turbine's position through the radially arranged catenary. The semi-submersible platform has a small draft and is stable during transportation and installation. The

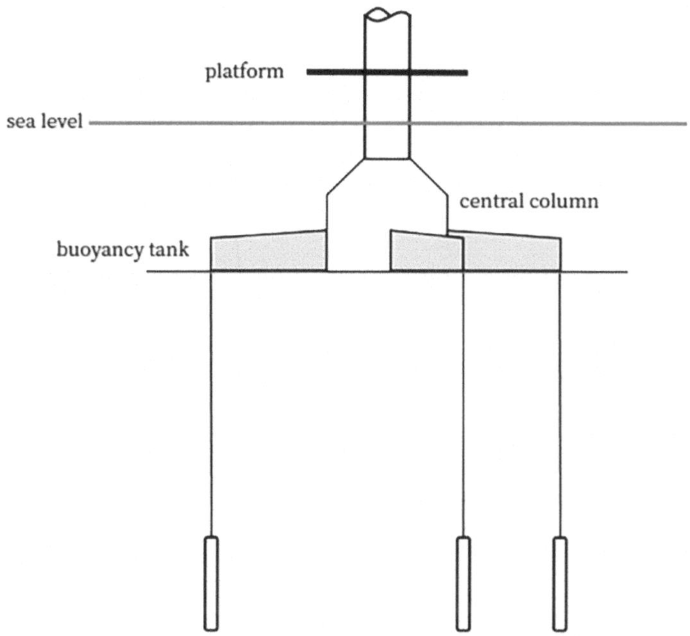

Fig. 3.12 Tension leg platform

corresponding cost is lower than that of the Spar-buoy and the tension leg platform. The semi-submersible platform structure is shown in Fig. 3.13.

3.5.2 Corrosion Prevention Measures for Offshore Wind Turbine and Booster Station Foundation

1. **Corrosive environment and zoning of offshore wind turbine and booster station foundations**

Offshore wind turbine and booster station foundations for a long time in harsh marine environmental conditions, by seawater immersion, salt spray erosion, waves, tidal wash, physical impact damage, and aquatic biological adhesion corrosion, etc., always facing the severe test of marine corrosion environment. These seriously affect the safe operation and service life of offshore wind turbines and booster stations. In different aquatic environments, the impact of various factors on corrosion is entirely different, such as oxygen content, humidity, pH salinity, etc., and offshore wind turbines and booster station foundation corrosion degree and corrosion rate differences are apparent. The typical marine environment can be divided into five different corrosion zones such as atmospheric zone, splash zone, tidal zone, submerged zone, and sea mud zone, as shown in Fig. 3.14.

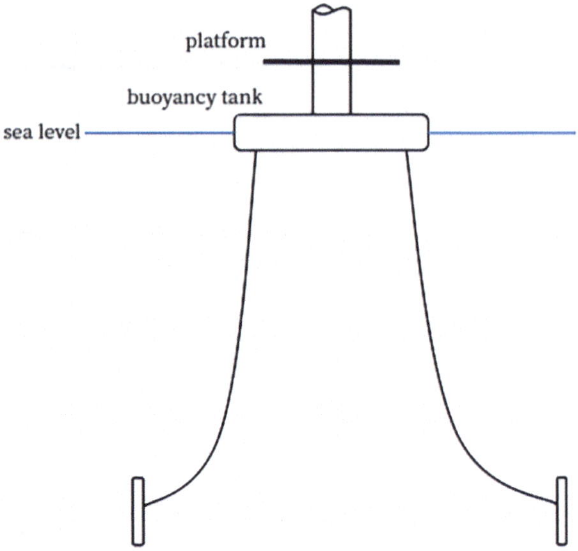

Fig. 3.13 Semi-submersible platform

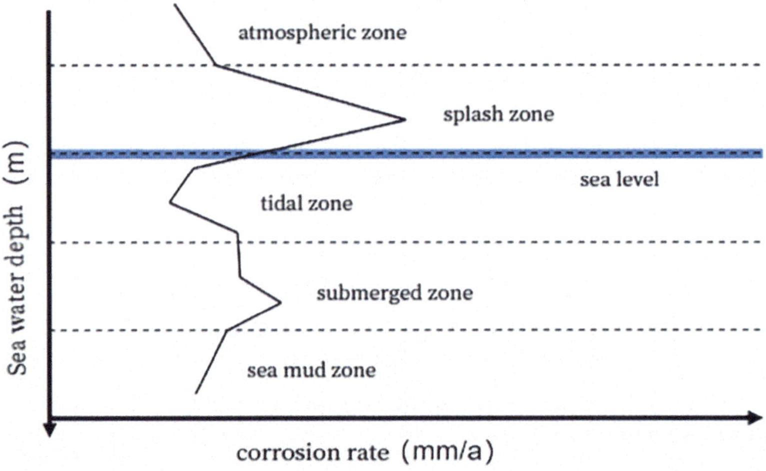

Fig. 3.14 Marine corrosive environment division

(1) Atmospheric zone

The marine atmospheric zone refers to the atmospheric zone above the splash zone of the sea surface and the coastal atmospheric zone, such as the area where the tower, nacelle, and blade are located. Relative to the ordinary inland atmosphere area, the effects of high humidity, high salt, high temperature, and dry and wet cycles are apparent. The corrosion medium in the marine atmosphere area is mainly salt spray, chloride ions, oxygen, and water vapor in the marine atmosphere. For the wind turbine steel structure exposed to the aquatic atmosphere, the long-term accumulation of corrosive media attached to the surface of the steel structure, the formation of an excellent liquid water film electrolyte, combined with the presence of a small number of carbon atoms in the composition of the steel structure, it is easy to form numerous primary cells, so that the surface of the steel structure to produce electrochemical corrosion and rust. Usually, the corrosion rate of materials in the marine atmosphere is 4 ~ 5 times higher than that in the land atmosphere [7]. Corrosion will not only damage the infrastructure of offshore wind turbines and booster stations, causing bolts and other fastening connections to reduce the strength, but also lead to a decline in the aerodynamic performance of offshore wind turbine blades, poor contact of electrical components, so that the failure rate of offshore wind turbines significantly increased, resulting in offshore wind turbine downtime due to failure, or even collapse.

(2) Splash zone

The splash zone refers to the section that can be wetted by the splash of waves above the average high tide level. In addition to corrosion factors such as sea salt content, humidity, and temperature, the splash area is also impacted by waves and soaked by tidewater. It has high sea salt particle content, long soaking time and frequent alternation between dry and wet. The splash area is the most severely eroded part in all marine environments for steel and reinforced concrete structures. Once severe local corrosion damage occurs in this area, offshore wind turbines and booster stations' foundation-bearing capacity will significantly increase. Reduce and shorten the service life, seriously affecting the safety and stability of offshore wind turbines and booster stations.

(3) Tidal zone

The tidal zone area refers to the area between the average high tide level and the average low tide level, which is submerged in water at high tide and exposed to the air at low tide; periodic changes characterize the alternating dry and wet. When submerged by high tide, it is affected by seawater corrosion, physical erosion, and cavitation corrosion, and when exposed to low tide, it is corrupted by the ocean atmosphere. In addition, the attachment and fouling of marine organisms have a more significant impact on the corrosion of steel structures, intensifying the corrosion of metal materials such as steel structures.

(4) Submerged zone

The submerged zone refers to the area below the perennial low tide line to the seabed. The wind power equipment in the seawater immersion area is completely submerged in seawater, such as the middle and lower parts of the foundation of offshore wind turbines and booster stations, which will be affected by factors such as dissolved oxygen, flow velocity, salinity, pollution, and marine organisms. Since the reduction reaction of oxygen controls the corrosion reaction of steel structures in seawater, dissolved oxygen plays a leading role in the corrosion of steel structures in the submerged zone [7].

(5) Sea mud zone

The sea mud zone below the seabed is mainly composed of submarine sediments. The seabed sediment's physical, chemical, and biological properties vary with the depth of the sea and seawater, and the corrosion environment is very complex. It has both the soil's corrosion characteristics and seawater's corrosion behavior, in addition to the influence of sulfate-reducing bacteria to accelerate the corrosion process. The sea mud zone contains less oxygen; the material corrosion process is slower than in other corrosion areas.

2. **Corrosion types and mechanisms of foundations of offshore wind turbines and booster stations**

The foundation structure of offshore wind turbines and booster stations is usually composed of reinforced concrete structures (gravity foundations) and steel structures (pile-supported foundations, floating foundations), easily eroded by the marine corrosive environment. According to the different construction materials of different foundation structures, the foundation corrosion types of offshore wind turbines and booster stations are mainly divided into two types: steel structure corrosion and concrete corrosion.

(1) Steel structure corrosion

Corrosion of steel structures is an electrochemical process; iron in steel structures is oxidized to a positive chemical valence state through electrochemical reactions in corrosive media. Corrosion of steel structures will cause the construction section to become smaller and the bearing capacity to decrease. According to the corrosion form, it can be divided into uniform corrosion and localized corrosion [8].

Uniform corrosion refers to the uniform corrosion damage of the contact parts of the steel structure and the medium so that the size of the steel structure becomes more minor and the color changes. Since the foundations of offshore wind turbines and booster stations are stable in various areas of the marine environment for a long time, different degrees of uniform corrosion are distributed on the entire surface of the foundation structure, resulting in a reduction in the thickness and strength of the components.

According to different corrosion conditions, mechanisms, and performance characteristics, it is mainly divided into pitting corrosion, galvanic corrosion, corrosion fatigue, filiform corrosion, waterline corrosion, impact corrosion, crevice corrosion, biological corrosion, and other forms. Localized corrosion means that only part of the steel structure in contact with the medium is damaged by corrosion, resulting in brittle damage to the steel structure and reduced durability. Local corrosion is more harmful than uniform corrosion.

1) *Pitting corrosion*

The appearance of small corrosion holes in localized areas of metal surfaces that develop in depth is called pitting. Pitting corrosion will accelerate the development of inward longitudinal depth, resulting in steel structure perforation, which is a very safety hazard and destructive. Mainly related to metallurgical factors, surface conditions, dispersion of salt particles, and atmospheric pollution.

2) *Galvanic corrosion*

Galvanic corrosion is a phenomenon in which dissimilar metals with different corrosion potentials contact the electrolyte to form a primary battery, accelerating the dissolution rate of metals with lower potentials and slowing down the dissolution rate of metals with higher potentials. The degree of corrosion depends on the presence of dissimilar metals. Relative differences in electric potential series in seawater.

3) *Corrosion fatigue*

Corrosion fatigue is a phenomenon in which the stress distribution of steel structures is uneven due to the joint action of cyclic stress or pulsating stress and corrosive media, which accelerates the formation of cracks. In a corrosive environment, the foundation of offshore wind turbines and booster stations must bear gravity loads and withstand alternating loads such as waves and storms. The combined effect of alternating load and corrosive environment will significantly reduce the fatigue resistance of steel structures [8].

4) *Filiform corrosion*

Filiform corrosion mainly occurs under metal coatings such as steel, aluminum, and magnesium. The corrosion head extends forward, leaving filamentous corrosion products. This usually happens at weak defects in the coating and at the edges and corners of components.

5) *Waterline corrosion*

Waterline corrosion occurs when the metal structure is semi-submerged due to the rich amount of dissolved oxygen. A corrosion line is first formed at the part slightly below the waterline.

6) *Erosion corrosion*

Erosion corrosion refers to the rapid erosion of the metal surface under the direct impact of high-speed water flow or high-speed fluid containing sediment particles and air bubbles. When suspended matter is present in seawater, the interaction between abrasion and corrosion is much more severe than when acting alone, with noticeable impact flow marks.

7) *Crevice corrosion*

Crevice corrosion occurs in cracks or gaps between metal parts, usually in the form of holes or corrosion spots, and the medium in the gaps is in a stagnant state. As the corrosion continues, the composition, concentration, pH value, etc. of the medium in the gaps The difference between the medium and the overall medium is getting bigger and bigger, resulting in a significant difference in the corrosion rate of the metal surface inside and outside the aperture, so that there are corrosion pits of different depths in the aperture. Crevice corrosion is mainly caused by unreasonable design or processing technology, sediment, scale, miscellaneous debris, rust layer, and biological deposition.

8) *Biocorrosion*

Biocorrosion occurs when organisms adhere to the steel structure, creating a new corrosive environment. Examples include uneven coverage of the metal surface, anaerobic bacteria activity, or hydrogen sulfide released by biological death and decay, which directly or indirectly promote metal corrosion. At the attachment site, a more obvious corrosion pit is formed.

(2) Concrete Structure Corrosion

In the marine environment, the main types of corrosion on concrete structures are chloride ion corrosion, carbonation, magnesium salt sulfate corrosion, alkali-aggregate reaction, and freeze–thaw damage.

1) *Chloride ion corrosion*

The high alkalinity of the hydrated cement produces a dense passivation film on the surface of the reinforcing steel within the concrete. This passivation film can only exist stably in a highly alkaline environment. The chloride ion in seawater is a highly penetrating corrosive substance that penetrates relatively quickly into the concrete and reaches the surface of the reinforcement. This decreases the pH value of the rebar surface, thus destroying the stability of the passivation film, forming a rust layer on the surface of the rebar, and corrosion will continue to develop on the inside.

2) *Carbonization*

The carbonization of concrete is the process by which the hydration products in cement stone interact with carbon dioxide in the environment to produce calcium carbonate or other substances. Although the products of the carbonation reaction will make the concrete dense in the short term, the surface concrete will become flimsy and brittle as carbonation proceeds. Under the washout of high-speed seawater waves, the surface concrete is highly susceptible to spalling, which reduces the thickness of the effective protective layer. The carbonation coefficient of concrete is not only related to factors such as air humidity, temperature, and CO_2 concentration but also to its water-cement ratio, material selection, and curing age.

3) *Magnesium salt sulfate erosion*

A sulfate attack is a common form of the chemical attack. Sulfate in seawater displaces $Ca(OH)_2$ in concrete to produce gypsum, which turns concrete into a paste or a non-adhesive substance. Gypsum is deposited and crystallized in the capillaries of cement stone, causing volume expansion, cracking the concrete, and destroying the protective layer of reinforcement. It also reacts with solid single sulfur type hydrated calcium sulfur aluminate and hydrated calcium aluminate to generate tri-sulfur type hydrated calcium sulfur aluminate containing a large amount of crystalline water, which produces local expansion force and makes the concrete structure swell and crack, resulting in a decrease in concrete strength and destroying the protective layer.

4) *Alkali-aggregate reaction*

Alkali-aggregate reaction refers to the process in which OH– in the concrete reacts with active SiO_2 in the aggregate to form a silica gel containing alkali metal. Silicone gel has a strong ability to absorb water and expand, causing uneven expansion of concrete, causing cracks, decreasing strength and elastic modulus, and affecting the durability of concrete [8].

5) *Freeze–thaw damage*

When the temperature of the saturated concrete structure in the cold area drops below the freezing point, the liquid water in the capillary pores of the concrete will freeze, and the volume will increase by about 9% when the water freezes, which will cause the expansion of the concrete. When the temperature rises under the sun's rays, the ice begins to melt, and when the temperature cools down again at night, freezing occurs again, causing further expansion. Freeze–thaw damage has a cumulative effect and may lead to concrete damage.

3. **Anti-corrosion measures for foundations of offshore wind turbines and booster stations**

The anti-corrosion measures for offshore wind turbines and booster station foundations are based on different construction materials. They are divided into anti-corrosion measures for steel structures and concrete materials.

(1) Anti-corrosion measures for steel structures

Corrosion protection of steel structures for the foundation of offshore wind turbines and booster stations usually adopts coating methods, plating methods, cathodic protection methods, reserved corrosion allowance methods, and selection of corrosion-resistant materials.

1) *Coating method*

The coating method belongs to isolation anti-corrosion, which is to spray anti-corrosion coating or paint coating on the surface of a steel structure to prevent various corrosive media such as water, oxygen, and chloride ions in the environment from penetrating the metal surface to prevent metal corrosion. The coating method is mainly suitable for marine atmospheric zones and splash zones. Commonly used anti-corrosion coatings include epoxy asphalt, rich epoxy, polyester coating, epoxy glass fiber reinforced plastic, etc. The auxiliary material is a curing agent, and the anti-corrosion period is 10 ~ 20 years, and the protection efficiency is 80–90%.

2) *Plating method*

The plating method is also classified as isolation corrosion protection and is mainly used in marine atmospheric, wave splash, and tidal zones. This method is used for most marine structures with small attached or connected parts.

3) *Cathodic protection*

The cathodic protection method belongs to electrochemical corrosion protection, which is divided into plus current cathodic protection and sacrificial anode cathodic protection. The former mainly uses high silicon cast iron anode material, the protected object as the cathode, under the influence of the external power supply, the formation of potential difference to prevent corrosion. The latter mainly uses zinc, aluminum, and other casting anode materials with higher activity than iron, which is welded to the structure to form a protective potential difference to stop the corrosion of steel structures.

4) *Reserved corrosion margin method*

In some environments, the medium's corrosion degree is not very high, and the material is not sensitive to the corrosion environment, so it is challenging to adopt conventional anti-corrosion methods. In this case, the technique of reserving corrosion allowance is often used in engineering to actively accept corrosion within a specific range.

5) *Selection of corrosion-resistant materials*

Corrosion-resistant iron and steel materials usually add certain alloying elements, such as chromium, phosphorus, and phosphorus, in the smelting of ordinary carbon steel to improve their corrosion resistance. It is necessary to fully consider the medium characteristics in the equipment design and manufacturing process, and the cost is relatively high.

(2) Anti-corrosion measures for concrete materials

The American Institute of Concrete (AIC) identifies four additional measures for adequately protecting reinforced concrete: epoxy-coated rebar, rebar inhibitors, cathodic protection, and concrete surface coatings.

1) *Epoxy coated rebar*

The powder made of filler, thermosetting epoxy resin, and admixtures such as cross-linking agent is sprayed on the preheated rebar treated on the surface by an electrostatic spraying process to form a certain brutal, impermeable, and continuous insulating coated rebar to prevent corrosion of rebar. Epoxy-coated rebar has the advantages of good adhesion to the base reinforcement, good tensile and flexural properties, low impact on concrete grip, good elasticity and friction resistance, alkali resistance, and chemical erosion resistance. The requirements for the production and construction process are very high to ensure the integrity of the epoxy coating on the surface of the reinforcement.

2) *Rust inhibitor for steel bars*

Rebar rust inhibitors can inhibit the electrochemical corrosion of rebar, which can effectively inhibit, stop, or delay steel corrosion by chloride ions. Rust inhibitors for rebar do not prevent harmful ions in the environment from entering the concrete but can turn off the harmful ions when they inevitably enter the concrete. Rebar rust inhibitor has the advantages of one-time use and long-term effectiveness, low cost of use, easy construction, and a wide range of applications. According to the principle of action, rust inhibitors can be divided into anode-type, cathode-type, and compound rust.

3) *Cathodic protection*

Cathodic protection is an effective auxiliary measure to reduce the corrosion rate of steel bars by using unique anode materials and using steel bars as cathodes to minimize the corrosion rate of steel bars by forming a potential difference after the corrosion of steel bars begins. Cathodic protection must strictly control the range of protection potential to prevent the reduction of concrete grip and hydrogen embrittlement caused by hydrogen evolution corrosion, and it should be used more carefully for prestressed concrete.

4) *Concrete surface coating protection*

Concrete surface coating protection is the application of organic coatings on the concrete surface to delay the corrosion rate of steel bars in concrete by isolating corrosive media. Concrete surface coating protection is more convenient and practical to protect reinforced concrete. It can effectively prevent the intrusion of corrosive media such as chlorides, soluble salts, oxygen, carbon dioxide, and seawater and fundamentally cut off the source of corrosion. The coating used should have Good alkali resistance, adhesion, and corrosion resistance, such as epoxy resin, polyurethane, acrylic resin, chlorinated rubber vinyl resin, and other coatings.

3.6 Technical Economic and Environmental Impact Analysis of Offshore Wind Farm

3.6.1 1Analysis of Investment Cost and Economic Benefit of Offshore Wind Farm

1. **Analysis of investment cost of offshore wind farm**

The investment cost of offshore wind farms mainly includes construction, ration, and maintenance costs. Among them, the construction cost of offshore wind farms is reflected in the depreciation of fixed assets throughout the wind farm's life cycle, mainly including equipment purchase costs, construction and installation engineering costs, other costs, and interest during the construction period. It is primarily affected by wind turbine equipment cost, installed capacity, seawater depth, offshore distance, and grid connection conditions. Under different construction conditions, the construction cost of offshore wind farms is different. Among them, the cost of equipment purchase (excluding the submarine cable in the collector line) accounts for about 50% of the construction cost, the cost of construction and installation engineering accounts for about 35% of the construction cost, the cost of sea-land use, project construction management cost, production preparation cost, etc. Other expenses account for about 10% of the construction cost, and the interest during the construction period accounts for about 5%. Among them, the interest during the construction period is related to the offshore wind farm's construction cycle and interest rate. With the continuous improvement of offshore wind power construction technology, especially the shortening of the project construction period, the interest will decrease to a certain extent. Offshore wind farm operation and maintenance costs are the changing costs of wind farms, including land lease, management costs, equipment failure maintenance costs, maintenance costs, spare parts purchase costs, insurance costs, financial costs, and other costs. The environment of offshore wind farms is harsh, and the operation and maintenance of wind turbines are relatively complex. The cost of operation and maintenance accounts for 20–30% of the total investment [9].

The investment and development costs of offshore wind farms are higher than those of onshore wind farms, mainly in the following aspects:

- Offshore wind farm projects are expensive upfront. The preliminary work takes a relatively long time, and it is necessary to coordinate with the local government, marine, and maritime departments to obtain supporting documents such as sea area, navigation, and marine environmental assessment.
- Offshore wind farm equipment purchase and construction and installation engineering costs are high. The environment of offshore wind farms is harsh, and the distance from the shore is long. Offshore wind turbines, offshore booster stations, submarine cables, foundations of offshore wind turbines and booster stations, and other equipment have high safety, reliability, and anti-corrosion requirements, which increases equipment manufacturing costs. Moreover, the cost and demand for submarine cables are much higher than land cables. On-site construction is complex, and the cycle is long. The requirements for installation ships and engineering experience are high, and the cost of construction and installation projects is relatively high.
- The operational and maintenance costs of offshore wind farms are high. The equipment that needs to be maintained for offshore wind farms mainly includes wind turbines, transformer substations and platforms, and submarine cables. Expensive vessels must be rented because offshore wind farms are generally located far from shore. In addition, adverse sea conditions such as typhoons and storm surges can cause large waves, making accessibility poor. Thus, the operation and maintenance of wind turbines are complex and costly.

2. **Economic Benefit Evaluation of Offshore Wind Farm**

The economic evaluation of offshore wind farms is similar to that of onshore wind farms and mainly includes financial and social benefit evaluation. According to the relevant regulations in the "Feasibility Study Report Compilation Regulations for Offshore Wind Farm Projects" (NB/T31032-2012) issued by the National Energy Administration, financial evaluation evaluates the project's essential survival ability, profitability, and sustainability from a project finance perspective within the current tax and pricing system for offshore wind power. It belongs to micro and direct benefit evaluation. Social benefit evaluation analyzes and evaluates the fundamental and long-term impacts of project construction and operation on local economic development, urban construction, labor employment, marine resource utilization, and the ecological environment from the overall perspective of the project and belongs to macro and indirect benefit evaluation.

The financial evaluation of offshore wind farms mainly includes the calculation of total costs (calculation of fixed assets value, total cost of offshore wind farms), calculation of power generation benefits (power generation revenue, taxes, profits, and distribution, etc.), solvency analysis (loan repayment of principal and interest) calculation, asset liability calculation), profitability analysis (financial cash flow calculation, capital financial cash flow calculation), financial viability analysis (financial plan cash flow calculation), and sensitivity analysis. This can analyze and demonstrate the financial feasibility and economics of offshore wind power investment projects and provide a solid basis for scientific decision-making of project investment. The evaluation of social benefits of offshore wind farms mainly lies in two aspects: energy-saving benefits and environmental benefits: energy-saving benefits are primarily reflected in the fact that no other conventional fossil energy is consumed during the operation of offshore wind farms; environmental benefits are mainly reflected in the fact that no pollution is emitted during the operation of offshore wind farms substances, harmful gases, and no need to consume water resources. Usually, offshore wind power is compared with coal-fired thermal power, and the saved coal consumption, pollutant discharge, and water saving are used as the energy-saving and environmental benefit evaluation indicators of offshore wind power.

3.6.2 Impact of Offshore Wind Farms on the Environment and Countermeasures

The specific construction and operation of offshore wind farms will inevitably impact the marine ecological environment, such as the impact on birds and fish, the impact on the water environment, the impact on shipping channels, and the impact on civilian and military facilities. It is necessary to take corresponding countermeasures to reduce or avoid the adverse effects of offshore wind farms on the ecological environment.

1. **Environmental impact analysis of offshore wind farms**

(1) Impact of offshore wind farms on birds

Offshore wind farms will affect the living environment of birds during construction and operation, mainly including impact, disturbance transfer, obstacle effect, habitat change, loss, etc. Among them, impact and interference transfer are the most concerning effects of international environmental protection scholars.

- Collision: If the offshore wind farm is located on birds' migration routes, the birds will collide with the wind turbine's rotor, tower, and nacelle and with auxiliary facilities such as power lines and wind measuring towers, causing fatal injuries or outright death.
- Disturbance transfer: Birds migrate to other areas to avoid threats and disturbances from offshore wind farms.
- Obstacle effect: To avoid offshore wind farms, birds change their migration or local flight paths.
- Habitat change and loss: The infrastructure of offshore wind farms includes generator bases, substations, submarine cables, etc., which may cause birds to change or lose their original habitats. The larger the offshore wind farm scale, the more pronounced the impact.

(2) Impact of offshore wind farms on fisheries

During construction, offshore wind farms can impact fishing activities and fishery resources. Most offshore wind farms are located offshore and will occupy limited fishing areas. The construction of offshore wind farms will prevent local trawling operations and interfere with fishermen's fishing activities and scope. The marine environmental waste pollution caused by the construction of offshore wind farms is not suitable for fish to eat; the accumulation of suspended solids causes the water body to lack oxygen, causing a large number of marine organisms to die; the increase in the concentration of particulate matter affects the vision of fish, making it impossible for them to eat food and nourish themselves Bad, eventually causing a lot of losses in fish farming.

(3) Impact of offshore wind farms on the water environment

During the construction of offshore wind farms, a large amount of domestic sewage will be discharged. If it is not treated in a timely and effective manner, it will cause damage to the surrounding environment. During the construction work, the piling and drilling of foundations of offshore wind turbines and booster stations and the laying of submarine cables require deep digging of trenches, resulting in the suspension of sediment and sediment on the seabed, making the water turbid, polluting the water quality of the sea area, and affecting the everyday life of marine organisms. After the offshore wind farm is built, the seabed environment will change due to artificial structures, changing the original sediment and hydrological characteristics, thereby affecting the diversity of benthic organisms and plankton.

(4) The impact of offshore wind farms on shipping channels

If the offshore wind farm site is close to the shipping channel, it may interfere with the ship's standard navigation. If it encounters severe weather, such as heavy fog and heavy rain, the boat may hit the wind power equipment, causing safety accidents and even casualties. During the operation of wind power equipment, a certain amount of electromagnetic radiation will be generated, which may affect the detection accuracy of radar, disorder the ship's communication function, and then affect the safety of navigation.

(5) Impact of offshore wind farms on civilian and military installations

Due to the Doppler effect generated by the blade rotation of offshore wind turbines and the radar cross-section effect of the wind turbine itself, it will interfere with air defense radars, civil aviation radars, submarine radars, military radars, etc., making radars ineffective in target detection and tracking, resulting in a fatal safety hazard. The rotation of the blades of offshore wind turbines will change the direction of rainfall, and the tower will block the transmission path of radar detection waves, increase the level of seismic noise, and affect the correct judgment of earthquake monitoring and nuclear explosion monitoring instruments.

2. **Measures to Improve the Marine Ecological Environment**

To reduce the adverse impact of offshore wind farms on the ecological environment, it is necessary to take relevant measures to improve the marine ecological environment.

(1) Scientific planning, rational layout

To reduce the impact of offshore wind farms on the marine ecological environment during construction and operation, scientific planning and rational layout should be implemented in the site selection stage of offshore wind farms to minimize possible adverse effects. First of all, the site of the offshore wind farm should be selected as far away from the habitat of marine organisms as possible and more from the deep sea to reduce the impact on marine ecology during the construction operation; secondly, it should be far away from sea lanes, ecological wetlands, and other areas, Minimize the interference to ocean transportation, which is conducive to further expanding the construction scale.

(2) Construction management, operation supervision

Construction management and operation supervision should be strengthened during offshore wind farms' construction and operation stages. First, the relevant departments must conscientiously perform their supervisory duties, carefully and strictly review the design plans of offshore wind farms, and never approve them if they do not meet the construction standards. After the construction of the offshore wind farm is completed, the supervision of its daily operation should be strengthened to ensure that it will not cause a severe impact on the surrounding environment. Secondly, the environmental protection department should maintain the ecological protection monitoring of offshore wind farms, accurately evaluate the water environment, fish, birds, fishery resources, and other living environments, and order rectification within a time limit for projects with environmental severe impacts, and try their best to maintain the balance of the marine ecosystem.

References

1. Chen X, Zhang X (2017) Offshore wind farm construction. China Electric Power Press, Beijing.

2. National Energy Administration (2019) NB/T10103–2018, Technical Specifications for Micro-siting of Wind Farms. China Water & Power Press, Beijing.
3. Pan W, Yang J, Sun F (2017) Wind power generation and grid integration technology. China Water & Power Press, Beijing.
4. Huang Z, Chen B, Zhou M (2016) Offshore wind power transmission engineering technology and application. China Water & Power Press, Beijing.
5. Wang Z (2013) Offshore wind power technology. China Machine Press, Beijing.
6. Li X, Liu Y (2016) Offshore wind farm construction guide. Hubei Science & Technology Press, Wuhan.
7. Wang J, Gao H, Liu B (2018) Offshore wind power anti-corrosion technology. China Electric Power Press, Beijing.
8. Chen D (2014) Offshore wind turbine foundation structure. China Water & Power Press, Beijing.
9. Feng Y, Chen X (2018) Offshore Wind Farm Economics and Risk Assessment. China Electric Power Press, Beijing.

Chapter 4
Offshore Wind Farm Construction

4.1 Introduction

The size and weight of offshore wind turbines (OWTs) are substantial, and the construction time window is limited by wind and wave conditions. Challenging maritime transportation and construction conditions, along with high technical difficulties, risks, and costs, impose strict requirements on the transportation capacity, operating water depth, and lifting capacity of the equipment. It is necessary to consider various factors affecting ocean hydrology, meteorology, navigation channels, and the seabed, and to design reasonable transportation and construction design schemes to ensure transportation and construction safety, reduce costs, and shorten the construction period.

This chapter mainly covers the transportation of offshore wind turbine equipment, the construction of offshore wind farms (OWFs), and the management of offshore wind farms.

4.2 Transportation of Offshore Wind Power Equipment

The transportation cost of offshore wind turbine equipment is an important part of the construction cost. The mode of transportation determines the construction efficiency and total cost to a large extent. Compared with the transportation of onshore wind power equipment, the size and weight of offshore wind power equipment components are larger, and the construction is more affected by the weather and equipment installation mode, so the difficulty and cost are doubled. To improve construction efficiency and reduce transportation costs, it is necessary to conduct detailed research on parts delivery, dock assembly, and maritime transportation to achieve the best strategy.

© The Author(s), under exclusive license to Springer Nature Singapore Pte Ltd. 2025
Y. Liu et al., *Offshore Wind Farm Technology*,
https://doi.org/10.1007/978-981-96-1889-7_4

4.2.1 Component Delivery

Component delivery involves transporting items such as towers, nacelles, hubs, blades, foundations, and booster stations from the supplier to the onshore assembly location or directly to the installation site at the port. Different components require different delivery methods. Offshore booster stations and wind turbine foundations are delivered directly to the installation site, eliminating the need for onshore assembly. Similarly, array and output cable-laying ships are pre-loaded with cables and do not require assembly at the port. Components like towers, nacelles, and blades should be delivered to the onshore assembly location at the port. Since these components are produced in various locations and are large in size, they must be transported to a dock for assembly before being loaded onto a ship. Water transportation is prioritized based on the supplier's location and component size; otherwise, land transportation is used, following requirements similar to onshore wind turbine transport. I is important to note that due to space limitations at the onshore assembly location and the conditions of offshore installation, not all components need to be delivered simultaneously. Deliveries should follow the established installation strategy, ensuring that components are supplied after the previous group is installed. To avoid storage issues, component suppliers can manufacture parts ahead of the planned delivery schedule.

4.2.2 Dock Assembly

Offshore conditions are challenging, with limited operational time and difficult installation processes, so the hoisting time of offshore wind turbines should be minimized. According to the installation strategy, the relevant components of the offshore wind turbine need to be assembled at the onshore assembly location at the port before being transported to the offshore wind farm by an assembly vessel. Therefore, during the construction of offshore wind farms, a dedicated dock is typically used for the temporary storage, assembly, and loading of components. However, there is currently a shortage of such specialized docks in China. As a result, some projects use large barges for simple assembly operations, while docks are primarily used for port collection and shipment.

The dock assembly method primarily depends on factors such as the hoisting process of offshore wind turbines, dock conditions, the type and capacity of construction vessels, and the location of the offshore wind farm. This method is categorized into the overall assembly of offshore wind turbines and the assembly of individual wind turbine components.

The overall assembly of offshore wind turbines involves assembling the towers, nacelles, and blades in suitable open spaces on the dock or on barges. This method requires a high hoisting capacity of the crane vessel for subsequent installation, along with overall lifting capability, stability during transportation, precise alignment of the assembled wind turbine with its foundation, and strong installation capacity.

The assembly of offshore wind turbine components is mainly categorized into tower assembly, three-blade assembly, rabbit-ear assembly, and others.

- Tower assembly involves installing three (or four) towers at the onshore assembly location, with the entire tower bolted to the deck of the assembly ship to maximize its load capacity.
- Three-blade assembly involves attaching the three blades and hub of an offshore wind turbine on land to form a wind rotor. This approach effectively reduces the amount of high-altitude work required, such as positioning and docking during offshore blade installation, thereby lowering the difficulty of offshore construction.
- Rabbit-ear assembly involves attaching two blades of an offshore wind turbine to the hub on land and then connecting them to the nacelle to make effective use of the deck area.

The main advantage of assembling offshore wind turbine components is that it reduces the need for a crane with high hoisting capacity during installation. However, it places high demands on ship stability during hoisting operations. Accurate alignment must be ensured between the lower and upper towers, the upper tower and the nacelle, the hub and the nacelle (in three-blade installation), or the third blade and the hub (in rabbit-ear installation). To achieve installation accuracy and maintain construction progress, platforms or ships with self-elevating pile legs are generally used.

4.2.3 Marine Transportation

Due to factors like natural environmental conditions, hoisting capacity, and ship performance, assembling offshore wind turbines can be challenging. To mitigate risks and shorten the construction period, the marine transportation method should align with the hoisting method for offshore wind turbines. Transportation is mainly divided into overall transportation and separate transportation.

(1) Overall transportation

Overall transportation involves transporting the fully assembled offshore wind turbine unit to the installation site. Given the large mass, high center of gravity, and significant windward area of the upper components (such as blades), there are stringent requirements for hoisting, loading, stability, and safety during transportation.

For large-scale offshore wind farms with many turbines, using large crane vessels for integrated transportation and installation may not meet the demands for transportation volume, intensity, and construction duration. Temporary installation methods are often cumbersome, time-consuming, labor-intensive, and economically inefficient due to their reliance on installation environment conditions. Therefore, a more effective approach is to use barges and transport boats for transportation and separate crane vessels for hoisting.

(2) Split transportation

Split transportation involves transporting offshore wind turbine components to the installation site after they have been assembled. Each individual component transported separately is significantly lighter than the entire assembled unit, making it easier to control stability and safety during shipment and transport. Most offshore wind farm projects, both domestically and internationally, use a split transportation scheme. This is because the specialized installation ships or self-elevating platforms used for offshore construction are designed or modified to accommodate split transportation. Typically, dedicated installation ships transport the components to a specific location before lifting begins. Once hoisting is complete, the ship moves to the next position. Components are generally transported by barges or other transport vessels.

In Europe, during the construction of offshore wind farms, installation vessels are typically used for transporting multiple sets of wind turbines at once. This method ensures stability and safety during lifting operations in open sea conditions. In China, where offshore wind power projects are generally closer to shore and self-elevating installation vessels are less common, transport barges are primarily used to move components between the dock and the installation vessel. However, with the increasing distance of offshore wind farms and the growing availability of self-elevating installation vessels in China, it is expected that using installation vessels for direct transportation will become a more prevalent trend.

During maritime transportation, it's essential to choose appropriate transport and pilot ships based on the number of offshore wind turbines and the parameters of the components. Before shipment, thoroughly understand the operation of lifting equipment, including slings and shackles, for hoisting and reshipping large components. When assembling deck cargo, avoid placing equipment like side scuttles, drain holes, and valves in areas that could interfere with crane operations. After loading the cargo onto the ship's deck, check the ship's stability. Typically, heavier cargo should be placed at the bottom of the hold to lower the ship's center of gravity, and cargo on the sides should be evenly distributed to prevent the ship from tilting.

4.3 Construction of Offshore Wind Farms

The construction of offshore wind farms is influenced by various factors such as weather conditions, ocean waves, engineering planning, technology, and management levels. Compared to onshore wind farms, offshore projects have longer construction periods, greater difficulty, and higher costs. Therefore, completing offshore wind farms on time and with high quality requires reasonable planning, scientific design, cost control, and risk management. Key aspects of offshore wind farm construction include the foundations, wind turbines, submarine cables, and offshore booster stations. A critical factor in the construction progress is the effective selection of professional installation ships. This section will focus on the construction process, the installation of offshore wind farms, and the role of specialized installation ships for offshore wind turbines.

4.3.1 Construction Process of Offshore Wind Farm

The process of offshore wind farm construction can be divided into four stages: planning and organization, manufacturing preparation, installation and construction, and testing and commissioning.

(1) Planning and organization stage

During the planning and organization stage, seabed exploration and testing should be conducted based on construction needs. Investigate construction conditions such as weather and wave information in the relevant sea areas, and establish construction time windows and navigation routes. Develop a comprehensive offshore wind turbine installation and construction plan tailored to the project's characteristics and requirements, and prepare advance plans to ensure the smooth progress of subsequent stages.

(2) Manufacturing preparation stage

In the manufacturing preparation stage, ensure the preparation and pre-assembly of equipment are completed while controlling project costs. This stage primarily includes the design, manufacturing, and construction planning of power transmission and transformation equipment; the preparation, planning, and pre-installation work for offshore wind turbine terminals; pre-installation tasks for towers and foundations; and the selection and preparation of wind turbine installation ships. These activities are crucial for shortening the construction period and controlling the costs of offshore wind power projects.

(3) Installation and construction stage

The installation and construction stage is the most critical, encompassing four main aspects: foundation construction, offshore wind turbine hoisting, offshore booster station construction, and submarine cable laying.

(4) Test joint commissioning stage

After completing the installation project, the next stage is testing and joint commissioning. This involves testing the operational performance and communication functions of individual offshore wind turbines, coordinating with the onshore monitoring center, conducting joint commissioning of the offshore wind farm, performing system grid connection testing, and evaluating power transmission and transformation capacity. These steps are essential to ensure the project is ready for formal and smooth operation.

4.3.2 Installation and Construction of Offshore Wind Farms

The installation and construction of offshore wind farms are critical phases and are significantly influenced by natural environmental factors such as sea conditions, wind conditions, and seabed geology. The processes and challenges involved in offshore wind farm construction differ markedly from those of onshore wind farms, particularly in areas such as foundation construction, offshore wind turbine hoisting, offshore booster station construction, and submarine cable laying.

4.3.2.1 Foundation Construction

Offshore wind turbine foundation types vary due to factors such as offshore distance, seawater depth, seabed geological conditions, wind turbine capacity, and construction resources. The construction processes for different foundation types differ significantly, with variations in required ships, equipment, time, and costs. The following will introduce the construction methods for three types of foundations: pile-supported foundations, gravity foundations, and floating foundations.

(1) Construction method of pile-supported foundation

Pile-supported foundations can be categorized into single-pile foundations, tripod foundations, jacket foundations, and pile-group foundations, based on the number of piles and their connection methods. Each type has distinct construction methods.

① *Construction method of single-pile foundations*

The single-pile foundation involves driving a large-diameter pipe pile directly into the seabed using a hydraulic hammer. This method is currently the most widely used for offshore wind farms. The construction process begins with determining the pile driving position. The pile is then erected vertically on the seafloor using a crane and grab device. A hydraulic hammer and pile gripper are used to drive the pile to the desired depth. During this process, sensors and detection equipment on the pile holder are used to ensure perpendicularity. Once the pile is securely fixed to the seabed, the transition piece is lifted and grouted into the pile. The time required for piling depends on factors such as seabed geological conditions, pile diameter and thickness, and the weight of the hydraulic hammer. If hard underground rock is encountered, drilling through the rock layer may be necessary using a drill bit inserted through the pile barrel.

② *Construction method of tripod and jacket foundation*

Tripod and jacket foundations are designed to better stabilize the structure, resulting in significantly smaller pile diameters and lengths compared to single-pile foundations. During the construction of tripod and jacket foundations, piles are either driven through the casing of each leg of the foundation or positioned in pre-drilled locations. The connection between the pile casings and the steel pipe piles is made underwater using methods such as pouring high-strength chemical slurry, filling with epoxy mortar, or underwater welding.

③ *Construction method of pile-group foundations*

The pile-group foundation construction method is relatively well-established and is currently used primarily in offshore wind power projects in China. This type of foundation consists of multiple small-diameter steel pipe piles (generally less than 2 m) and pile caps. The use of these smaller piles means that the requirements for piling equipment are lower and construction is relatively straightforward. However, the process involves reinforcing concrete cushion caps with steel cages, which requires binding, concreting, and curing at sea. This leads to a longer construction period and higher manufacturing and construction costs.

(2) Gravity foundation construction method

Gravity foundations offer advantages such as a simple structure, ease of construction, and a lower cost compared to single-pile foundations. They are stable and less affected by seabed conditions. Typically, reinforced concrete gravity foundations are prefabricated on land and then transported to the site for installation. The process begins with an excavation vessel digging a deep pit on the seabed and ensuring the pit bottom is level. The gravity foundation is then lifted and lowered into the pit. Sea sand is pumped into the hollow foundation to settle and compact, after which excess water is pumped out. Due to their massive size and weight, gravity foundations require large barges and floating cranes for installation. Additionally, they demand a suitable seabed and are challenging to level at sea. They are well-suited for areas with solid natural foundations but are less appropriate for soft foundations or areas prone to seabed scour.

(3) Construction method of floating foundation

Floating foundations are designed for deep-sea wind farms with water depths greater than 50 m. These foundations are typically prefabricated and assembled on land before being towed to their designated locations by ships. Alternatively, the entire foundation and wind turbine unit can be dragged to the installation site after assembly at the dock. Due to their complex design, high dependence on favorable weather conditions during construction, and the current stage of development for operational control strategies, floating foundations are still in the model or prototype testing phase both domestically and internationally.

4.3.2.2 Hoisting of Offshore Wind Turbines

Hoisting offshore wind turbines is a critical aspect of offshore wind farm construction. The hoisting processes vary greatly depending on factors such as installation process requirements, the capacity of the installation vessel, and site conditions. Generally, hoisting methods are classified into two categories: integral hoisting and split hoisting.

For integral hoisting, the dock is selected as the assembly site. After the main components of the offshore wind turbine unit are assembled and commissioned at the dock, the entire unit is transported to the installation site. At the site, a crane vessel hoists the wind turbine onto the platform. Anchor positioning is crucial to avoid collisions between the tower and the platform. Additionally, a soft-landing system with special devices and an overall translation and alignment system are needed to ensure precise alignment between the tower and the platform.

However, integral hoisting requires high-quality professional assembly docks and favorable weather conditions, making it less commonly used in the industry. Split hoisting, on the other hand, is more prevalent due to its flexibility and the availability of lifting resources. The split hoisting method, similar to land-based operations, includes single-blade unit lifting, rabbit-ear unit lifting, and three-blade unit lifting methods.

(1) Unit hoisting method of the single-blade

The single-blade lifting method involves hoisting the three blades of an offshore wind turbine separately. It is mainly divided into two types: Y-shape lifting and horizontal lifting. Y-shape lifting does not require an adaptive design for wind turbine cranking, but it involves a complex blade installation hoisting structure and has lower reliability. Horizontal lifting allows for faster installation but poses challenges for direct-drive offshore wind turbines, particularly with cranking the blades. There are three main lifting methods for the single-blade unit.

① The first type of lifting method involves handling the tower, nacelle, generator, hub, and blades individually. Initially, the tower is lifted section by section using one or two cranes. Next, the nacelle, generator, and hub are lifted separately, followed by the individual lifting of each blade. This method does not require pre-assembly of components onshore and has lower demands on dock resources, with relatively lighter lifting weights. However, it necessitates the design of supporting tooling, lifting locations, and fixtures for each component. This method is characterized by long offshore lifting times, inefficiency, higher costs, and a greater risk of interruptions, making it less suitable for areas far from the coast. The lifting vessel can be either a self-elevating platform barge or a self-propelled installation vessel.

② The second type of lifting method involves pre-assembly of the tower, nacelle, and generator, with the hub and blades lifted separately. In this method, the nacelle and generator are assembled at the dock. Initially, the tower is lifted section by section using one or two cranes. Next, the assembled nacelle and generator are hoisted together, followed by the lifting of the hub. Finally, each blade is lifted individually. This method involves the pre-assembly of a few components onshore and has relatively low dock resource requirements. Since the nacelle and generator are lifted simultaneously, the crane's lifting capacity needs to be higher than in the first method. This requires designing and developing assembly support tooling and single-blade hoisting fixtures for the nacelle and generator. Additionally, the wind turbine unit must have designated lifting points and appropriate slings for the nacelle and generator, resulting in lower lifting efficiency and higher costs. The lifting vessel for this method can be a self-elevating platform barge.

③ The third type of lifting method involves pre-assembling the nacelle, generator, and hub, with the tower and blades lifted separately. Initially, the nacelle, generator, and hub are assembled at the dock. Next, the tower is lifted section by section using one or two cranes, followed by the lifting of the pre-assembled nacelle, generator, and hub. Finally, each blade is lifted individually. This method involves pre-assembly of components onshore, avoiding the need for onshore impeller assembly and requiring less dock resource. However, it still necessitates the design and development of assembly support tooling and single-blade lifting fixtures for the nacelle, generator, and hub. The wind turbine units must also have specific lifting points and corresponding slings for these components, leading to moderate lifting efficiency and costs. Self-elevating platform barges are suitable for this lifting method.

(2) Unit lifting method of rabbit-ear unit

The rabbit-ear lifting method involves pre-assembling two blades of the offshore wind turbine onto the hub before lifting. There are two main approaches for this lifting method:

① The first type of rabbit-ear lifting involves pre-assembling the hub with two blades to form a rabbit-ear impeller before lifting. The process starts with assembling the hub and two blades at the dock to create the rabbit-ear impeller. The tower is then lifted individually, section by section, using one or two cranes. Following this, the nacelle, generator, and the assembled rabbit-ear impeller are lifted separately. Finally, the third blade is lifted vertically. This method requires pre-assembly of components onshore and has general requirements for dock resources. However, it demands a minimum crane lifting capacity of around 300 t at the hub height and necessitates the design of assembly support tooling and vertical blade lifting fixtures. Additionally, the wind turbine unit must be equipped with lifting points and tooling slings for the rabbit-ear assembly, resulting in average lifting efficiency and costs. Suitable lifting vessels for this approach include self-elevating platform barges, self-propelled installation vessels, or crane vessels.

② The second type of rabbit-ear lifting involves pre-assembling the nacelle, generator, hub, and two blades onshore to create a rabbit-ear-type head. The process begins with assembling these components at the dock. Subsequently, the tower is lifted in sections using one or two cranes, and the assembled rabbit-ear-type head is then hoisted. Finally, the third blade is lifted separately. This method requires extensive pre-assembly of onshore components and places high demands on dock resources. The lifting crane must have a significantly higher capacity at the hub height, generally not less than 600 t, compared to the first method. It also requires the design and development of specialized assembly support fixtures for the rabbit-ear-type head and vertical blade lifting. Additionally, the wind turbine units must be equipped with suitable lifting points and tooling hangers for the rabbit-ear-type head assembly and lifting. This method offers high lifting efficiency and average lifting costs. Suitable lifting vessels for this approach include self-elevating platform barges, self-propelled installation vessels, or heavy lifting vessels.

(3) Unit hoisting method of three-blade

The unit lifting method for the three-blade assembly involves pre-assembling the three blades and the hub on land to form a wind wheel, which is not initially connected to the nacelle. During maritime transportation, the angle of the blades must be adjusted to optimize deck space usage. For offshore installation, the nacelle is first hoisted onto the tower, followed by the direct hoisting of the pre-assembled wind wheel onto the nacelle. This method streamlines the offshore installation process by minimizing the steps required for blade positioning and docking, thereby reducing construction difficulty and time. There are three main unit lifting methods for the three-blade assembly.

① The first method involves pre-assembling the tower, nacelle, generator, hub, and three blades onshore. Initially, the tower is lifted in sections by one or two cranes. Next, the nacelle and generator are hoisted separately. Finally, the assembled wind wheel is lifted using both main and auxiliary cranes. This approach requires significant onshore assembly time and places high demands on dock facilities due to the large size of the assembled wind wheel. The minimum lifting capacity needed for the crane at the hub height is typically no less than 300 t, and specialized tooling for the wind wheel assembly is necessary. Additionally, the wind turbine unit must be equipped with specific lifting points and appropriate lifting equipment. This method offers slightly higher lifting efficiency and incurs general lifting costs. Self-elevating platform barges are commonly used as lifting vessels.

② The second method involves pre-assembling the tower, nacelle, generator, hub, and three blades onshore. Initially, the tower is lifted in sections by one or two cranes. Next, the pre-assembled nacelle and generator are hoisted. Finally, the assembled wind wheel is lifted using both main and auxiliary cranes. This approach necessitates extensive onshore pre-assembly and places high demands on dock facilities. The nacelle and generator are lifted simultaneously, and the assembled wind wheel is large, requiring a crane with a minimum lifting capacity of no less than 400 t at the hub height. Specialized tooling for both nacelle and generator assembly and wind wheel assembly is needed. Additionally, the wind turbine unit must have lifting points and tools for both the nacelle and generator as well as the wind wheel. This method offers high lifting efficiency and reasonable costs. Self-elevating platform barges or self-propelled installation vessels can be used for lifting.

③ The third method involves pre-assembling the tower and the nacelle, generator, hub, and three blades onshore. The assembly of the tower takes place at the dock, while the nacelle, generator, hub, and blades are pre-assembled separately. The process begins with the integral lifting of the assembled tower. Next, the pre-assembled nacelle and generator are lifted, followed by the hoisting of the wind wheel. This approach requires large-scale onshore pre-assembly and demands a high standard for dock facilities, including a foundation platform for tower assembly. The lifting cranes must have a minimum capacity of around 800 t. Additionally, specialized tooling is required for nacelle and generator assembly, as well as for wind wheel assembly. The wind turbine unit needs lifting points and corresponding tools for these components. Although this method offers high lifting efficiency, the cost of onshore assembly is substantial. Self-elevating platform barges can be used as lifting vessels.

4.3.2.3 Construction of Offshore Booster Station

The offshore booster station is a key component of the power transmission and transformation system of large offshore wind farms, which is used to increase the output voltage of offshore wind turbines (such as 35 kV) to a higher level of 110 kV/ 220 kV for long-distance transmission. The construction of offshore booster stations mainly includes foundation construction and upper platform construction.

(1) Foundation construction

The construction of offshore booster stations involves various foundation types, each with distinct processes. Commonly used foundations include gravity, single-pile, and jacket foundations. Since the construction methods for these foundations are similar to those used for wind turbines, they will not be detailed here.

(2) Upper platform construction

To minimize on-site installations, avoid potential quality issues from on-site welding, and shorten the installation and commissioning time of offshore equipment, it is recommended to use onshore assembly for the upper platform of offshore booster stations when large floating cranes are available. In this approach, each layer of the structure is prefabricated and assembled onshore, including the installation of electrical equipment, forming a complete superstructure that can be shipped as a whole.

The construction of the upper platform involves several key steps: loading the upper module, selecting the appropriate transport vessel, planning the lifting scheme, and performing the offshore installation. The lifting scheme is crucial and must address aspects such as layered lifting points, the combination of single layers into the overall structure, and the setup of upper hangers. Offshore installation methods for the booster station topside are primarily the floating method and the crane boat lifting method. The floating method adjusts the water volume and tide level of the ballast tank to sink the transport ship, allowing the topside to be installed onto the foundation as a whole. This method is suitable for oversized and heavy modules but requires precise matching of transport vessel size and foundation design, and is currently not used in China. The crane boat lifting method, which involves lifting the entire topside from the transport ship and installing it on the foundation, is simpler and more commonly used.

4.3.2.4 Submarine Cable Laying

Submarine cables are categorized into internal array cables and output cables based on their applications. Internal array cables connect offshore wind turbines to each other and to offshore booster stations, while output cables transmit the electrical energy generated by offshore wind farms to the power grid. The primary distinction between the two is their voltage levels, but their laying construction processes are quite similar.

The submarine cable laying plan is developed based on factors such as hydrogeological conditions, offshore distance, and installed capacity. The process, which follows a generally consistent sequence, includes construction preparation (such as layout of traction optical cables and sea sweeping), cable transportation, initial landing construction, cable laying and protection in the sea, installation and fixation of cables at offshore wind turbines or booster stations, and final testing and acceptance.

(1) Construction preparation

An effective and cost-efficient submarine cable laying scheme should be developed based on the micro-siting of offshore wind farms, including the landing points and routing of the cables. This involves selecting appropriate optical cables and construction methods, and using tugboats to tow sea sweeping anchors to remove remaining seabed obstacles, such as fishing nets.

(2) Cable loading and transportation

The installation site is typically the dock of the submarine cable manufacturer. During loading, the construction ship is moored and secured. Cable trestles are employed to transport the submarine cables to the ship, where they are reeled onto the fixed cable cabin or cable reeling platform. Cables, stored on pallets or spools, are lifted directly onto the ship's deck using a crane.

(3) Top landing construction

According to the specific situation of the sea area, the landing shore should be located with flat terrain and moderate water depth. The submarine cable laying ship should be close to the shore as possible, and the anchor boat shall anchor and locate. Use the landing point winch to roll back the cable and pull the submarine cable to the set position at the landing point. When landing the submarine cable, it is necessary to use the foam buoy to bind the buoy at a certain interval of the cable so that the submarine cable can surface when landing.

(4) Laying and protection of the middle piece of cable

Submarine cables are typically protected either by direct burial or external covering. Direct burial protection methods include punching and burying, knife plowing, cutting, and pre-excavation. External covering protection involves using artificial covers to shield the cables.

① *Direct burial protection method*

Flushing and burying method: The excavation section using a buried plow is equipped with a nozzle, and the high-pressure water sprayed is flushed through the seabed. After exiting the trench, guide the submarine cable and interrupter to the trench through a dedicated guide rail or channel. The burying plow is mainly divided into shoe-type punching and burying machines fixed to the ship's side and plow-type burying machines. The operating water depth is generally less than 60 m, and the maximum buried depth can reach 10 m. The burying speed is related to the seabed geological conditions and buried depth, generally ranging from 1 to 15 m/min.

- The knife plowing method involves a buried plow towed by a cable-laying boat using a lead rope. The knife plow at the rear of the plow creates a trench on the seabed, into which the submarine cable and interrupter are embedded. This method employs larger cable-laying vessels that require fewer auxiliary ships and offer strong wind resistance, speeding up the construction process. These vessels are equipped with condition monitoring and control systems. The burial depth is adjustable via the control system, with the latest technology allowing for depths of over 4 m and operational water depths of up to 2000 m.
- The cutting method involves using cutting wheels, chains, or other specialized mechanical shredders to dig trenches along the planned route before laying submarine cables. This method is suitable for seabed rock or hard soil layers. It is characterized by slow construction speed and higher costs, making it generally used for short-distance cable laying.
- The pre-excavation method involves several steps: first, a backhoe dredger is used to excavate the trench; then, a submarine cable laying vessel places the cable into the trench; and finally, a dredger fills in the trench. This method is suitable for areas with thick seabed silt layers but involves many construction procedures, resulting in slower construction speed and higher costs.

② *Coverage Protection method*

The cover protection method involves using various manual coverings, such as half-cast iron pipes, concrete slabs, cement fillers, and riprap blocks, to safeguard submarine cables that have already been laid. This method is primarily used in areas where avoiding rocky sections is not feasible, where cables intersect with other pipelines, where the seabed soil layer is too thin, or where there are restrictions on large vessels. Among these methods, riprap is relatively common, employing specialized ships to transport materials and flexible rock-fall pipes to distribute rubble along the submarine cable route. Careful control and monitoring are essential throughout the laying process.

(5) Installation and fixation of submarine cables on the side of offshore wind turbines or booster stations

First, pull the lead wire out of the J-tube and connect it to the submarine cable that needs to be installed. Then, slowly lower the submarine cable to the seabed. Next, gradually pull the lead wire from the upper end of the J-tube, guiding the submarine cable towards the inlet at the lower end of the J-tube until it enters the J-tube. Finally, continue to pull the lead wire until the submarine cable is fully pulled through and connected to the offshore wind turbine or booster station for fixation.

(6) Test and acceptance

Before a submarine cable is put into operation, a DC withstand voltage test must be conducted to evaluate its insulation performance and ability to withstand overvoltage. This test effectively identifies local defects, such as mechanical damage or moisture in the cable insulation. The basic principle of the DC withstand voltage test is to apply a direct current voltage higher than the normal operating voltage to the main insulation of the submarine cable, maintaining this voltage for a certain period while keeping it constant. If the cable does not exhibit any breakdown during the test, it is considered to meet the operational requirements and can be safely put into use.

4.3.3 Professional Installation Vessels for Offshore Wind Farms

4.3.3.1 Classification of Professional Installation Vessels for Offshore Wind Farms

Professional installation vessels are essential for constructing offshore wind farms. Depending on their functions and purposes, these vessels mainly include crane ships, self-propelled and self-elevating installation vessels, jack-up platform barges, heavy lifting vessels, cable-laying vessels, and feeder vessels.

(1) Crane ship

A crane ship, also known as a floating gondola, is a self-propelled vessel equipped with jack-up legs to create a stable lifting platform. It is primarily used for loading and unloading oversized cargo at ports or anchorages, as shown in Fig. 4.1.

Crane ships are mainly categorized into fixed and rotary types, depending on the lifting equipment they carry. The boom of a fixed crane ship is stationary in one direction, similar to a mast crane—it can only lift and lower but cannot rotate. To achieve rotation, the entire ship is slowly turned by tugboats or by pulling anchor chains, allowing the cargo to move with the ship to reach the desired position. In contrast, the boom of a rotary crane ship can rotate 360°, making its mechanical structure significantly more complex than that of a fixed crane ship. Due to its draft, displacement, and working depth, the rotary crane ship is often used in deep-sea wind

Fig. 4.1 Crane ship

farm construction. With a large deck area and high loading capacity, it effectively reduces the time required for transporting parts back and forth, thereby lowering transportation costs. Additionally, it offers greater lifting capacity and height, making it ideal for the installation of large-capacity wind turbines.

(2) Self-propelled and self-elevating installation vessel

A self-propelled and self-elevating installation vessel is a large barge with 4–6 legs, capable of traveling at speeds of 8–12 nmile/h and supporting a deck load capacity of 1500–6500 t, as shown in Fig. 4.2. The primary difference between this type of vessel and a self-elevating platform barge lies in its self-propulsion capability and lifting capacity. With its self-propelled ability and maneuverability, the vessel offers greater flexibility. Its larger deck capacity and space can typically accommodate 6–8 offshore wind turbines, reducing reliance on ports. Equipped with large cranes and piling equipment specifically designed for the installation of offshore wind turbines, the vessel can quickly reposition itself, elevate stably and reliably, and withstand harsh sea conditions and weather. This extends the operational window and enhances installation efficiency.

(3) Jack-up platform barge

A jack-up platform barge is typically equipped with a crane and 4–8 legs, as shown in Fig. 4.3. Upon arrival at the site, the legs are lowered into the seafloor to secure and stabilize the vessel. By adjusting the hydraulic lifting system, the barge can be raised completely or partially out of the water, creating a stable platform unaffected by wave motion. The size of a jack-up platform barge falls between that of a crane ship and a self-propelled, self-elevating installation vessel. Small jack-up barges can generally carry 2 offshore wind turbines, while larger ones can accommodate 6–8 turbines. Unlike self-propelled vessels, jack-up platform barges rely on hawsers for mooring and require tugboats for transportation, with speeds typically ranging from 4 to 8 knots. The transition process is time-consuming, making operation less convenient and dependent on relatively calm sea conditions.

Fig. 4.2 Self-propelled and self-elevating installation vessel

Fig. 4.3 Jack-up platform barge

Fig. 4.4 Heavy lifting vessel

(4) Heavy lifting vessel

The heavy lifting vessel is designed with a high-capacity crane and lacks an additional lifting system, featuring a barge-like shape, as shown in Fig. 4.4. These vessels include types such as tripod cranes, jib cranes, and other floating cranes. While they are commonly used in offshore oil and gas industry construction, they are less frequently employed for offshore wind turbine installation. However, heavy lifting vessels can be utilized for tasks such as foundation construction, overall hoisting of offshore wind turbines, and the construction of booster stations.

(5) Cable laying ship

The cable laying ship, also known as a cable-laying vessel, is primarily used for installing internal array cables and output cables in offshore wind farms and can also serve as a cable maintenance vessel, as shown in Fig. 4.5. This vessel can be either a large barge or a self-propelled barge specifically designed for cable laying operations. In shallow waters, positioning methods typically include anchoring and dynamic positioning. The mooring positioning method adjusts the ship's displacement by controlling the length of the anchor chain. This method is highly reliable, easy to operate, adaptable to various environmental conditions, and cost-effective, making it valuable for engineering applications. During cable laying, considerations include the vessel's motion response to wind and waves, the mooring system's capacity to handle environmental loads, and the forces exerted by submarine cables and stingers on the hull.

(6) Expansion ship

Expansion ships are essential at every stage of offshore wind farm installation and include crew boats, utility boats, tugboats, diving support boats, and dredging/scouring vessels. Their main roles during installation are to support the activities of the installation ship, such as crew transportation, anchoring, and towing, and to transport components, particularly foundations, from the shore base to the offshore wind farm. The size and composition of expansion ships depend on factors such as the specifications of the installation vessel, transportation strategy, distance from the shore,

Fig. 4.5 Cable laying ship

environmental conditions, the weight of offshore wind turbines and foundations, project scale, chartering costs, and vessel availability.

4.3.3.2 Selection of Professional Installation Vessels for Offshore Wind Farms

The choice of professional installation vessels for offshore wind farms significantly affects both construction costs and progress. It is essential to consider a range of factors, including project scale, construction conditions, offshore distance, component specifications and weight, installation strategies, and associated costs. Key technical parameters to evaluate include maximum lifting capacity, maximum operating height, minimum operating radius, crane slewing capacity, number of cranes, maximum load capacity, deck area, draft, navigation capability, positioning capability, leg lifting capability, and other facility features.

(1) Maximum lifting capacity

The maximum lifting capacity refers to the highest rated load that the crane can safely lift under normal working conditions. For cranes with variable amplitude, this capacity is defined at the minimum amplitude. The maximum lifting capacity of the crane determines the installation vessel's capability, ensuring it can handle at least the maximum weight of the object being hoisted. When selecting an installation vessel, it is important to consider not only the maximum load but also factors such as the operating radius, the height of the wind turbine, and the crane's load curve.

(2) Maximum operating height

The maximum operating height refers to the highest point from the center of the boom hook to the deck within the crane's minimum operating radius, and it is typically related to the length of the crane's boom. Given that the center hub of offshore wind turbines is generally positioned at a considerable height, there are stringent requirements for both lifting weight and height. Therefore, when selecting a crane, it is essential to consider the crane's load curve and choose an installation vessel with appropriate lifting capacity and working height.

(3) Minimum operating radius

The minimum operating radius is the distance from the center of the crane to the vertical line of the hook when the crane is raised to its maximum height. This radius is typically influenced by the crane's positioning. When hoisting a wind turbine, it usually refers to the shortest distance from the center of the crane to the center of the wind turbine. This distance must be less than the distance from the crane's center of rotation to the nearest edge of the deck. During wind turbine installation, the crane's operating radius is determined by the crane's position on the vessel and the distance between the installation vessel and the wind turbine.

Lifting capacity, operating height, and operating radius are interrelated and mutually restrictive. This relationship is typically illustrated by the lifting performance curve or corresponding data tables. When selecting an installation vessel, it is crucial to ensure that these three factors meet the installation requirements of the offshore wind farm.

(4) Crane slewing capacity

Cranes installed on ships are generally categorized into two types: full-slewing cranes and fixed-arm cranes. A full-slewing crane can rotate its boom 360° and can be positioned and angled freely, with no specific site requirements for installation. In contrast, a fixed-arm crane can only lift objects directly in front of it and has stringent requirements for the installation angle. During hoisting, adjustments to the crane's position are limited to climbing or repositioning methods.

(5) Number of cranes

The number of cranes refers to the total cranes available for assembly and hoisting operations on the installation vessel. Typically, an installation vessel is equipped with a main crane for primary hoisting tasks and an auxiliary crane for assembly, transfer, deck layout, and tail sliding. The number of cranes significantly affects the hoisting method, progress, and organizational plan of the hoisting operations.

(6) Maximum load

The maximum load refers to the highest weight that the vessel is permitted to carry. This capacity reflects the vessel's loading ability, transportation capability, hull size, and overall survivability. When selecting the maximum load, factors such as the distance from the shore, transportation strategy, installation strategy, and the specifications and quality of the components must be considered.

(7) Deck area

The steel plate arranged horizontally on the hull is known as the deck. As a critical component of the hull, the deck supports various cabins and mechanical equipment and is vital for maintaining the vessel's strength and unsinkability. The deck area influences the vessel's loading capacity, transportation capacity, assembly capability, and overall construction organization plan. In the installation of offshore wind turbines, vessels with larger deck areas are preferred as they can effectively address transportation challenges, reduce the number of required ships, and lower transportation costs.

(8) Draft depth

Draft depth is categorized into design draft depth and structural draft depth, with the latter being greater. The design draft depth is the vertical distance from the lower edge of the plate keel to the upper edge of the design waterline in the midsection of the hull. This depth is crucial for ensuring the vessel's safety and stability during operation. In contrast, the structural draft depth is the vertical distance from the lower edge of the plate keel to the design load line, ensuring the vessel's structural integrity during operation. The draft depth (excluding Jack-up vessels) is influenced by both navigation and construction depths. For offshore wind farm installation vessels, the draft typically ranges from 6 to 10 m.

(9) Navigational ability

Navigation capability refers to the ship's ability to operate autonomously. An installation vessel without self-propulsion typically requires a tugboat or booster device for movement during transfers. The speed of a tugboat usually ranges from 4 to 5 knots, while a booster device generally operates at 2.5–4 knots. In contrast, a self-propelled installation vessel, depending on its power, can typically achieve speeds of 7–8 knots. Navigation capability impacts the vessel's maneuverability, transfer efficiency, need for additional towing assistance, installation costs, planning, and selection of hoisting positions.

(10) Positioning ability and leg lifting ability

The positioning capability of an installation vessel is primarily determined by its dynamic positioning system, which influences the mooring speed and method for installing wind turbines. A vessel equipped with a dynamic positioning system can swiftly and accurately reach the intended installation site. For jack-up installation vessels, the lifting capacity of the legs depends on factors such as the number and style of legs, the lifting technology employed, and the lifting system's method. The lifting capacity of the pile legs affects the time required for pile insertion, the insertion depth, and the workable water depth.

(11) Other facility settings

Other facilities on the installation vessel, such as personnel living areas, logistics support facilities, and helicopter platforms, while not decisive, play a crucial role in the wind turbine installation process. A larger living area allows more personnel to be accommodated, which can expedite construction. Well-equipped logistics support facilities enhance worker morale and contribute to smoother operations. Additionally, a helicopter platform provides maneuverability and flexibility for emergency rescues and personnel transfers.

4.4 Construction Management of Offshore Wind Farm

The construction management of offshore wind farms shares similarities with onshore wind farms, focusing on controlling and managing construction costs, quality, scheduling, and safety. However, construction scheduling and safety management are particularly critical, as they significantly impact project costs and quality. Offshore wind farm construction faces unique challenges compared to onshore projects, such as limited weather windows, harsh ocean conditions, and specialized transportation and installation strategies. These factors contribute to longer construction periods, increased difficulty, and higher costs. Consequently, this section will emphasize the management of construction scheduling and safety for offshore wind farms.

4.4.1 Construction Scheduling Management of Offshore Wind Farms

4.4.1.1 Construction Scheduling Management Method of Offshore Wind Farm

Construction scheduling management for offshore wind farms involves the owner regularly inspecting the construction unit's progress reports or the supervision unit tracking and providing feedback on the actual site conditions.

(1) Confirm the project scheduling reporting system

The supervisory engineer requires the contractor to submit the overall scheduling plan before issuing the construction order. Additionally, the contractor must provide the scheduling plan for the following month the day before the monthly regular meeting, the weekly scheduling plan the day before the weekly regular meeting, and a daily report on construction progress from the previous day. The supervision department will then prepare and submit a daily supervision report to the owner, detailing the construction progress and engineering quantities completed the previous day, as reviewed by the supervisors.

(2) Check the scheduling of the project regularly

① *Onsite inspection*

The project supervision engineer should conduct daily inspections of the construction site to monitor progress, record site activities, and require the contractor to provide daily progress records. These records should include the amount of work completed and accumulated each day, the number of personnel and machinery actively involved, production efficiency, any idle personnel and machinery, the presence of the contractor's supervisors and technicians, and any specific reasons affecting project progress.

② *Submit monthly scheduling report*

The supervisory engineer should require the contractor to submit a detailed monthly scheduling report on project completion. This report should include an overview or general description, project progress, project photos, financial status, and any other relevant matters.

(3) Coordinate scheduling progress

The supervisory engineer should require the contractor to submit a detailed monthly scheduling report on project completion. This report should include an overview or general description, project progress, project photos, financial status, and any other relevant matters.

(4) Submit monthly project scheduling report

The supervision engineer submits the project scheduling report to the owner as part of the monthly supervision report, focusing on issues affecting the project schedule and providing improvement recommendations.

(5) Establish a scheduling control account

The supervision engineer prepares a detailed first-level network plan based on the project milestone plan provided by the owner. They use a software system to analyze daily deviations between the actual construction progress and the contractor's updated progress plan. Before each regular project meeting, the engineer reviews the progress of the previous project stage, analyzes the current progress plan, and provides recommendations. They may also request the contractor to accelerate progress and adjust the progress plan as needed.

4.4.1.2 Factors Influencing Construction Scheduling of Offshore Wind Farms

The construction scheduling of offshore wind farms is influenced by various factors, including the enterprise's decision approval process, procurement management procedures, marine environmental conditions, seabed geological conditions, equipment supply capacity, construction vessel requirements, and external social factors.

(1) Enterprise decision approval process

Unified centralized control within enterprises can lead to a cumbersome and slow project decision approval process. While centralized management helps prevent regulatory violations, it reduces flexibility and autonomy in the project. This lack of flexibility can hinder overall construction progress.

(2) Enterprise procurement management procedure

If the enterprise adopts a centralized procurement strategy, the group company will manage procurement activities with a unified plan. All items requiring purchase must be included in the procurement plan, and procurement will strictly adhere to this plan. Items not listed cannot be procured. While centralized procurement reduces procurement violations, its lack of flexibility and lengthy processes can negatively impact project construction, leading to delays in construction scheduling.

(3) Marine natural environment conditions

During construction, offshore wind farms are highly susceptible to extreme weather and natural conditions, such as typhoons, high waves, tidal currents, sea ice, and sea fog. Adverse weather can lead to accidents such as offshore construction vessel capsizing, fractures in wind turbine towers or blades, foundation damage, and even casualties among construction personnel. These issues can significantly disrupt construction progress and increase costs.

(4) Seabed geological conditions

Offshore wind farms involve a lengthy process from site selection and planning to construction. During this time, seabed conditions can change due to ocean current erosion, potentially rendering the original design plan unsuitable. The accuracy of seabed geological surveys can be affected by marine environmental constraints and the performance of sea-sweeping and drilling equipment, leading to discrepancies from the initial design. These deviations are often discovered during construction, necessitating adjustments to the original design and planning, which can impact the construction schedule.

(5) Equipment supply ability

The construction of offshore wind farms involves various components, each requiring different equipment and installation strategies. The timeliness and reliability of equipment manufacturers in delivering products and services directly impact the construction schedule of offshore wind farms.

(6) Equipment requirements for construction ships

The construction of offshore wind farms requires specialized vessels, such as crane ships, self-propelled and self-elevating installation vessels, self-elevating platform barges, heavy lifting vessels, cable-laying ships, and support vessels. As competition in offshore wind power development intensifies, there is a shortage of these professional ships. The limited number and performance of available vessels often fail to meet the optimal requirements for offshore wind farm construction, leading to delays or suspensions in construction schedules.

(7) External social environment

The complex external social environment encompasses the political, ecological, and economic contexts. Key factors include government agencies and their staff, power grids, local residents, and businesses. The attitudes and responses of these stakeholders towards offshore wind farms can significantly influence construction scheduling.

4.4.1.3 Improvement Measures for Construction Scheduling of Offshore Wind Farms

To achieve the construction scheduling targets for offshore wind farms, it is essential to evaluate the key factors affecting scheduling in advance and develop an improvement plan to mitigate or reduce their impact. Considering these factors, the main measures to enhance the construction scheduling of offshore wind farms are as follows:

- Simplify the enterprise decision approval process, optimize procurement management procedures, and reduce the initial preparation time for offshore wind farm construction.
- Coordinate the scheduling of all relevant units and ensure adequate maneuvering time to prevent accidents.
- Monitor weather conditions closely, adjust construction schedules to accommodate weather windows, halt operations during sudden adverse weather, and implement protective measures to prevent major safety incidents and economic losses.
- Focus on the service capacity of equipment suppliers and implement measures to control equipment supply scheduling, ensuring that construction timelines are not affected by equipment shortages or quality issues.
- Based on the construction needs of the offshore wind farm, promptly determine the selection of professional construction vessels and finalize service contracts with the vessel providers as soon as possible.
- Strengthen the review of design plans and drawings, and ensure that designers promptly update and modify any plans and drawings that do not align with actual conditions.
- Promptly urge the construction team to make necessary adjustments based on actual conditions and expedite the construction schedule.
- Engage in communication and consultation with local government, enterprises, and residents to minimize or avoid the impact of external social factors on the construction schedule.

4.4.2 Construction Safety Management of Offshore Wind Farm

4.4.2.1 Basic Regulations on Construction Safety Management of Offshore Wind Farms

Construction safety management is a crucial aspect of offshore wind farm construction. The fundamental requirements for construction safety are outlined in the national industry standard NB/T 10393-2020, "Technical Code for Construction Safety of Offshore Wind Farm." These provisions include:

- Establish and Improve Safety Management Organization: The construction unit should set up a production safety committee and establish a comprehensive safety management system. Responsibilities for safety management should be clearly defined, with a focus on developing and implementing safety rules and regulations. This ensures that production safety is standardized and effectively managed throughout the engineering construction process.
- Implement Comprehensive Safety Supervision and Inspection Mechanisms: The construction unit should establish a robust system for supervising, inspecting, and investigating potential safety hazards. This system must be actively applied throughout the entire construction site process to ensure effective safety management.
- Establish an Emergency Management System: The construction unit should develop a comprehensive emergency management system and detailed emergency plans for various types of safety incidents. This includes coordinating with survey, design, construction, and drill units to create specific emergency plans. Implement emergency organizations, procedures, resources, and measures, and conduct regular drills to ensure the effective execution of emergency responses.
- Establish Safety Management Structures and Goals: Each participating unit should set up a dedicated safety management organization and enforce a safety production responsibility system. They should define overall and annual safety production goals, ensure adequate safety investments, and maintain high standards of on-site construction practices.
- Personnel Qualifications: Construction site personnel must possess the fundamental knowledge and skills required for their specific tasks. Offshore operators, special operators, and those handling specialized equipment must complete specialized safety training and pass relevant examinations. They should only be permitted to work after obtaining the necessary qualifications.
- Technical and Safety Disclosures: The survey and design units should provide detailed technical and safety disclosures to all participating units prior to the commencement of construction. This ensures that all parties are fully informed about the project's technical requirements and safety protocols.
- Construction Permit and Planning: The construction unit must apply for a construction permit in compliance with relevant regulations. Prior to commencing construction, it should conduct on-site surveys, identify potential hazards, and prepare detailed construction organization designs, construction plans, and safety technical measures. For high-risk projects, specialized construction plans should be developed. For large-scale or particularly hazardous projects, experts should be engaged to review and approve the special construction plans.
- Safety Signage: The construction unit should plan and set up safety signs at the construction site. Whenever the construction environment or operation procedures change, they must reassess the hazardous and harmful factors on-site and update the safety signs accordingly.
- Emergency Plans: The construction unit should develop an emergency rescue plan and an on-site disposal plan tailored to the project's characteristics and scope. They

should also monitor areas and sectors of the construction site that are prone to accidents

- Regular Safety Inspections: The construction unit should regularly conduct on-site safety inspections and investigate and address hidden dangers. They must strictly enforce safety measures at the construction site and prevent illegal commands, unauthorized operations, and violations of labor discipline.
- Safety Technical Disclosure: Based on the characteristics of the project construction operations, potential dangers and harmful factors, and the relevant safety operating procedures or technical measures, the construction unit should provide a safety technical disclosure to the construction personnel and complete the necessary signing procedures.
- Supervisory Unit Responsibilities: The supervisory unit should carry out supervision in compliance with laws, regulations, and mandatory standards related to project construction. It is responsible for overseeing construction project safety and ensuring adherence to these requirements.

4.4.2.2 Comprehensive Safety Management of Offshore Wind Farm Construction

According to the national industry standard NB/T10393-2020 "Technical Code for Construction Safety of Offshore Wind Farm Engineering," comprehensive safety management in offshore wind farm construction encompasses various aspects, including offshore operators, construction temporary facilities, construction electricity, transportation, fire prevention and explosion protection, ship operation, work at height, welding operations, diving operations, seasonal and special environmental construction, ship machinery and equipment, safety protection measures, and individual protective equipment. This chapter focuses on the relevant management requirements specifically for offshore operators, construction temporary facilities, transportation, ship operation, diving operations, and seasonal and special environmental construction.

(1) Offshore operators

- Offshore operators should possess the necessary physical and psychological fitness for the job. They must be aware of the hazardous and harmful factors present on offshore construction sites and understand the corresponding preventive measures and emergency response procedures.
- Offshore operators must not consume alcohol before or during their time at sea. They should not work alone without supervision and are prohibited from swimming or fishing while offshore.
- During offshore operations, operators must correctly wear personal protective equipment and use the appropriate labor protection tools. Construction personnel should avoid entering dangerous areas when not actively working.

(2) Construction temporary facilities

- The onshore transfer base for offshore wind farm projects should be located close to the wind farm site and must meet requirements for flood control, moisture prevention, and typhoon protection.
- Temporary docks should be situated in open water areas with stable bank slopes, low waves, and flow velocities. They should have suitable water depths, good geological conditions, and convenient land transportation access. Safety warning signs must be installed.
- The construction operation area of a large construction ship should be divided into safe channels, ensuring that no obstacles are placed within these channels.
- The construction site of the offshore operation platform should prioritize the safety of construction personnel by setting up safety warning signs, protective facilities, and life-saving equipment.
- Temporary navigation aids should be installed according to design requirements and can only be removed once the permanent navigation markers have been completed and passed acceptance inspection.
- The width of the temporary pedestrian springboard at sea should not be less than 60 m. It must have sufficient strength and stiffness to meet requirements. The springboard should be equipped with safety guardrails or safety nets, securely fixed or tied at its end, and the surface should include anti-slip facilities.
- The water quality of the water supply at the construction site should meet the requirements, and the water supply pipelines in cold and severe cold areas should have insulation and antifreeze measures.
- Protective measures should be implemented for dangerous areas and parts of the construction site, with clear and prominent safety warning signs placed to alert personnel.
- Construction vessels, offshore operating platforms, and land bases should be equipped with communication equipment, such as radios and satellite phones, that meet radio communication standards and the International Convention for the Safety of Life at Sea.

(3) Transportation

- The construction unit should keep detailed records of personnel boarding the ship and going to sea, including their names, ages, affiliations, boarding times, departure and arrival times, and contact numbers.
- Ships should display lights or shapes according to regulations. When sailing, crew members must avoid approaching the sides of the ship that do not have safety guardrails.
- The construction unit should develop a dispatch and tugboat plan as well as an emergency plan before deploying the ship. The construction ship should be sealed and reinforced accordingly. During tugboat operations, ensure smooth communication, record weather and wave conditions, and have designated personnel monitor and document the navigation status of the towed ship, including lights and waterline marks. If it becomes necessary to anchor for

shelter from the wind, the ship should follow regulations and either enter port or anchor appropriately.

- Maritime transportation should meet the relevant requirements such as traffic ship navigation, large-scale equipment and facilities transportation, tugboat, untying cable operation, anchor operation, ship ladder collection and release, personnel barging and boarding, ship navigation in bad environments, and so on. For details, please refer to the national industry standard NB/T10393-2020 "Safety Technical Code for Offshore Wind Farm Engineering".

(4) Ship operation

- The ship's operational performance must be suited to the working conditions of the sea area. This includes ensuring that the vessel is capable of handling the specific environmental and operational challenges encountered in that region.
- In the event of severe weather conditions, such as strong winds, heavy fog, thunderstorms, or storms, the construction ship should cease operations and evacuate personnel to a safe area.
- During nighttime operations, the construction ship should be equipped with sufficient lighting facilities, warning lights, and signal signs to ensure visibility and safety.
- Ship operations should adhere to the relevant requirements for hoisting and outboard operations, as outlined in the national industry standard NB/T10393-2020 "Technical Code for Construction Safety of Offshore Wind Farm Engineering."

(5) Diving operation

- The qualifications of divers must comply with the current industry standard JT/T955 "Qualification Conditions for Divers."
- The diving operation site should be equipped with a first aid kit and corresponding first aid equipment. For operations where the water depth exceeds 30 m, decompression chambers and other necessary equipment must also be provided.
- When the water temperature is below 5 °C, the flow velocity exceeds 1 m/s, or there are hazardous organisms, obstacles, or pollutants in the construction waters, appropriate safety protection measures should be implemented.
- Diving operations should include the use of special control ropes, diving telephones, and supply lines. Divers must use dedicated diving ladders, which should be securely attached to the diving boat.
- When delivering tools, materials, and items to divers, they should be conveyed using ropes rather than being thrown directly into the water.
- Divers should use specialized tools when installing components underwater to ensure that the components are securely positioned. Additionally, the air supply pipe should not be placed between the components or operated against the water flow direction

- Divers should remove any obstacles and sharp objects from the casing before entering a large-diameter casing. Additionally, the water level inside the casing should be higher than the water level outside the casing.

(6) Seasonal and special environment construction

- During offshore construction, anchor positions for construction ships should be arranged based on the prevailing wind directions of the monsoon to ensure stability and safety.
- Before the thunderstorm season begins, check the lightning protection devices for lifting, piling, and other equipment on the construction site.
- During construction in high-temperature seasons, ensure timely distribution of heatstroke prevention and cooling items, adjust working hours reasonably, and implement adequate ventilation and cooling measures.
- During winter construction, implement anti-skid measures for on-site roads, offshore platforms, stairs, scaffolding, and ship decks. Clear ice and snow before work begins, and apply anti-freezing measures to foam fire extinguishers, oil–water pipelines, and lifeboat lifting devices on the ship's deck.
- During the monsoon period, extend the anchor cable of the construction ship as needed. Adjust the ship's position in response to strong wind, high waves, and current pressure. Ensure that watertight facilities for doors, windows, hatches, and holes are in good condition and that the drainage system is clear. Equip equipment such as pile frames, lifting jibs, bridges, hooks, pile hammers, and cranes with proper sealing devices.
- The construction unit should develop a typhoon prevention emergency plan. Before a typhoon, move the ship to a sheltered anchorage in advance, and unload material-laden ships as soon as possible. If unloading cannot be completed on time, adjust and seal the ship's balance. Install temporary protective ropes and guardrails on both sides of the deck and pedestrian walkways.

4.4.2.3 Emergency Rescue for Offshore Wind Farm Construction

Offshore wind farm construction emergency rescue is a related emergency plan to be adopted in case of safety accidents during the construction process. The relevant regulations on emergency rescue in the national industry standard NB/T10393-2020 "Technical Code for Construction Safety of Offshore Wind Farm Engineering" are as follows:

- Maritime operators should be proficient in survival skills at sea, as well as in rescue and first aid procedures for electric shock incidents.
- The construction unit should enhance the maintenance and management of emergency materials and equipment to ensure the effective functioning of the emergency management system. Additionally, it should establish a mutual assistance agreement for emergency rescue with neighboring construction units.

- Offshore construction sites should be equipped with emergency vessels to handle any crises that arise.
- The construction unit should display an emergency handling flowchart in a prominent location, such as the bridge of the construction vessel.
- In the event of an emergency at an offshore operation site or aboard a ship, reporting and handling should adhere to the following requirements:

  The site supervisor or the person in charge of the nearby vessel should immediately notify the offshore construction area manager and sound the alarm to alert authorities. The offshore construction area manager should then promptly report to the project safety committee and, if necessary, to the maritime authorities, marine sector, maritime police, and other relevant departments for rescue coordination.

  Immediately activate the emergency plan, swiftly control the sources of the hazard, notify neighboring construction units to assist in the rescue efforts, and organize the rescue of those in distress. Implement emergency rescue measures or arrange for the evacuation of on-site personnel based on the severity of the accident.

  Request neighboring vessels to assist in the rescue as needed.

  Implement necessary measures to prevent the escalation of the accident and the occurrence of secondary or related disasters. Maintain order at the accident scene and protect the scene and related evidence.

Chapter 5
Offshore Wind Power Forecasting

5.1 Introduction

Offshore wind power forecasting is the necessary foundation for the operation and control management of offshore wind farms and the new energy power system. It directly affects the power generation indicators and accuracy assessment rewards and penalties of wind farms and determines the core competitiveness of wind farms participating in the electricity market. High-precision forecasting of offshore wind power can help reduce the cost of power generation over its entire life cycle. Furthermore, it is an important means to mitigate the adverse effect of wind power volatility on the power system and increase the proportion of offshore wind power grid connections.

Compared to onshore wind power, the weather conditions in offshore wind farms are more complex and harsher, making high-precision forecasting more difficult. Besides, the increasing strictness of forecasting accuracy assessment, the intensification of penalties, and the accelerated implementation of the electricity market pose numerous challenges to offshore wind power forecasting.

The rapid development of technologies such as cloud computing and artificial intelligence has brought new opportunities for the development of offshore wind power forecasting technology.

This chapter first summarizes the characteristics of offshore wind power and the unique challenges it brings to power forecasting, then summarizes the classification and significance of offshore wind power forecasting, and finally introduces several core cutting-edge technologies for offshore wind power forecasting.

© The Author(s), under exclusive license to Springer Nature Singapore Pte Ltd. 2025

Y. Liu et al., *Offshore Wind Farm Technology*,
https://doi.org/10.1007/978-981-96-1889-7_5

5.2 Challenges for Offshore Wind Power Forecasting

The complexity of weather, extreme wind conditions, and complex internal flow patterns of offshore wind farms lead to the low predictability of offshore wind power; the forecasting is far more difficult than that of most onshore wind farms, mainly reflected in the following five aspects.

(1) Offshore weather is complex and diverse, evolving rapidly. Meteorological factors such as temperature, air pressure, and wind speed are prone to sudden changes, and the occurrence frequency of ramp events increases obviously. In addition, the temporality characteristics of wind speed series are complex and challenging to extract and learn, which directly limits the accuracy of forecasting models based on the extrapolation principle of time series, including ultra-short-term forecasting.
(2) Also, due to the complex coupling effect of the atmosphere, ocean, and waves, the offshore numerical weather prediction model needs to consider more complex interactions of coupled mode variables, making high-precision forecasting more difficult and directly limiting the accuracy of short-term wind power forecasting.
(3) Offshore wind farms have a large development scale and a smooth underlying surface, with most wind turbines centrally located and arranged in a regular pattern, leading to more apparent wake effects; the spatial and temporal connections between wind resources and output are closer and more complex. On the one hand, this increases the uncertainty and modeling complexity of power forecasting; on the other hand, it exacerbates the superposition effect of power aggregation and power fluctuations, increasing the forecasting risk.
(4) Extreme weather at sea, such as typhoons, is more frequent than on land, and the ramp events are significant. Studies have shown that the fluctuation of wind farm output can be around 60% of the rated power under 15 ~ 20 min scale and 90% of the rated power under an hour scale. Extreme weather can cause widespread shutdowns of offshore wind farms, resulting in serious power drop events and huge impacts on the grid. Therefore, the forecasting demand for typhoons and ramp events is more significant.
(5) Compared with onshore wind power, offshore wind power forecasting requires a wider time scale and a finer time resolution. On the one hand, the accessibility of offshore wind farms is poor, and the operation and maintenance (O&M) window is shorter than that of onshore wind farms. Medium and long-term wind power forecasting is essential to support the operation and maintenance plans of offshore wind farms. On the other hand, real-time forecasting at second and minute time scales is required to support wake control and active grid support control in offshore wind farms.

5.3 Classification and Significance of Offshore Wind Power Forecasting

Overseas research on wind power forecasting technology began in the 1980s. Relevant technology development and system applications in China started in 2008. With the rapid development of China's wind power industry, in 2011, the "Notice of the National Energy Administration on the Issuance of Interim Measures for the Management of Wind Farm Power Forecasting" required that all wind farms operating on the grid must establish a wind power forecasting system and transmit data to the grid dispatching agencies in real-time. Early wind power forecast assessments were limited to ultra-short-term forecasting from 15 min to 4 h in advance and short-term forecasting from 24 to 48 h in advance. In recent years, the period required for wind power forecasting by the grid dispatching agencies have gradually increased, with short-term forecasting up to 1–10 days in advance. However, due to the unique operating environment of offshore wind farms and the high-risk nature of offshore wind projects, offshore wind farms' operational control and maintenance management also places higher and newer demands on wind power forecasting, reflected in the wider temporal-spatial scale, finer resolution, and higher accuracy and reliability requirements. It makes the importance of offshore wind power forecasting under multiple temporal-spatial scales and its uncertainty analysis more prominent.

Therefore, this chapter elaborates on various forecasting techniques in terms of time scale and expression methods, including the definitions, significances, main technical routes, and technical difficulties. According to the time span (or time scale), wind power forecasting can be classified into real-time forecasting, ultra-short-term forecasting, short-term forecasting, and medium and long-term forecasting, involving scales from seconds and minutes to daily, monthly, and yearly. According to the expression of forecasting results, it can be divided into deterministic forecasting (or single-point forecasting) and uncertainty forecasting (including probabilistic forecasting, interval forecasting, scenario forecasting, etc.).

5.3.1 Divided by Time Scale

5.3.1.1 Real-Time Forecasting

At present, the definition of offshore wind speed or wind power real-time forecasting has not yet formed a unified standard, and it is generally accepted that real-time forecasting should focus on higher temporal resolutions, such as seconds and minutes. The International Energy Agency (IEA) Wind Tasks 32 (Wind Lidar) and 36 (Wind Forecasting) working groups define the real-time forecasting range of offshore wind farms as 1 s to 1 min in advance.

The implementation of real-time forecasting is mainly based on time series algorithms. The high temporal resolution wind data mainly derives from upstream flow observations by wind turbine nacelle-based Lidar, wind turbine SCADA data, and offshore wind tower data. The main application scenarios including:

(1) Feed-forward control of wind turbines.
(2) Whole-field operation control of wind farms oriented towards generation increase and load reduction, such as full-field wake control.
(3) Station and group control serving active grid support, coordinated management with energy storage systems, etc.

Real-time forecasting mainly adopts the time series modeling approach; the biggest difference with existing forecasts is the higher requirements for time resolution. The higher time resolution brings two challenges to real-time forecasting: on the one hand, the volatility and randomness of wind speed and power time series increase significantly (see Fig. 5.1), making it more challenging to extract temporal features; on the other hand, more accurate extraction of temporal features and higher forecasting accuracy often rely on the more complex model, but the complex models require more computation time, which conflicts with the time-validity of high time resolution forecasting.

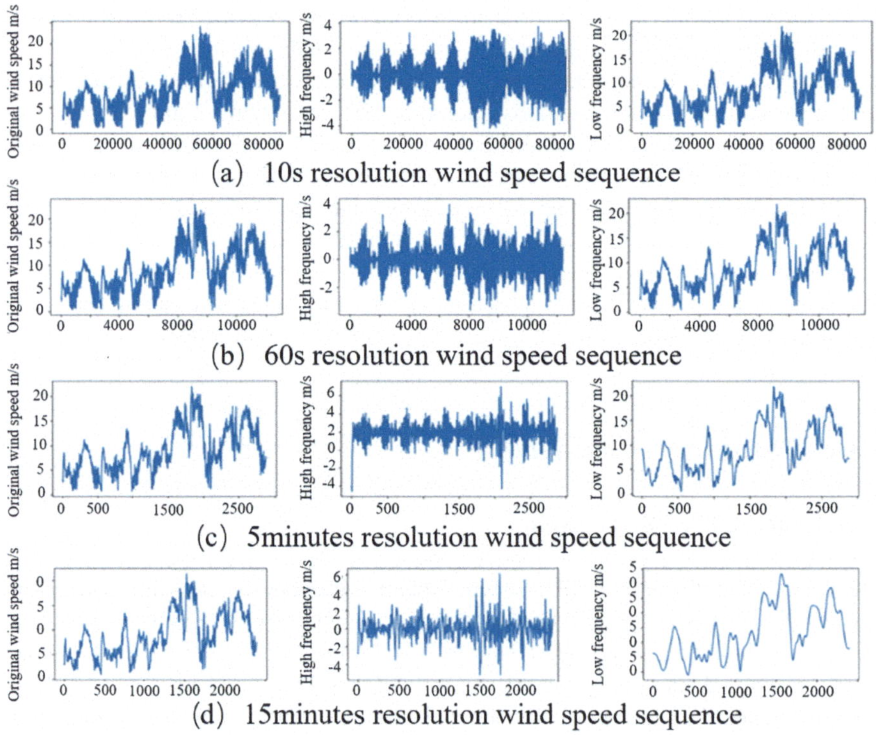

Fig. 5.1 Frequency domain information of wind speed series at different time resolutions

5.3.1.2 Ultra-Short-Term Forecasting

According to IEC standards, "Interim Measures for the Management of Wind Farm Power Forecasting", and "Implementation Rules for Coordinated Operation of Wind Power Forecasting and Power Grids", the time scale of ultra-short-term forecasting is from 15 min to 4 h in advance, with a time resolution of 15 min. Similar to real-time forecasting, ultra-short-term forecasting is mainly achieved through time series methods and typically uses historical wind speed observation sequences (or power sequences) to extrapolate future wind speed sequences (or power sequences). The historical observation data includes wind measurement towers, LiDAR, wind turbine SCADA, etc. Due to the time series extrapolation principle, the accuracy of the ultra-short-term forecasting decays significantly over time, making it difficult to directly use the extrapolation technique for short-term power forecasting with a longer period, the Numerical Weather Prediction (NWP) is needed.

Ultra-short-term offshore wind power forecasting encounters similar problems to real-time forecasting, i.e., the stronger volatility of wind speed/power time series makes the extraction of time series features more difficult, resulting in reduced model accuracy. However, compared to real-time forecasting, ultra-short-term forecasting has lower time-validity requirements, so current research focuses on two aspects.

(1) Improving the deep neural network structure to enhance the network's ability to extract temporal and spatial features and improve forecasting performance.
(2) Adopting the refined modeling route, using clustering algorithms to pre-classify offshore weather and output scenarios first, and then developing different modeling/error correction methods under different scenarios to improve forecasting accuracy.

5.3.1.3 Short-Term Forecasting

Similar to ultra-short-term forecasting, the definition and assessment criteria of short-term wind power forecasting are clear both domestically and internationally, i.e., forecasting 24–72 h in advance (some provinces or regions will extend it to 7 or 10 days). NWP is the core input data for short-term wind power forecasting, with other observations as secondary input data. The future wind power is obtained by establishing the mapping relationship between NWP data and output power.

Short-term wind power forecasting results are the main reference information for the power dispatching agencies to formulate the day-ahead dispatching plan of the power system and are important for optimizing wind turbine combinations, economic dispatching, and standby optimization. For the main players in the electricity market, short-term wind power forecasting is the basis for wind power participation in the spot market, and the forecasting accuracy significantly impacts the electricity price and the effectiveness of participating players. In addition, for offshore wind farm operators, short-term forecasting can guide short-term operation and maintenance plans, avoiding power losses due to downtime during high wind periods. At the same time, the forecasting results can also be used to estimate and pre-judge the offshore accessibility and optimize the operation and maintenance strategies of offshore wind farms.

5.3.1.4 Medium and Long-Term Forecasting

Medium and long-term forecasting usually focuses on future power generation under a week, month, or annual scale, with a time resolution of hours or days. Similar to real-time forecasting, medium and long-term forecasting of offshore wind power is an emerging research hotspot both domestically and internationally, with no unified definition of the forecasting time scale.

Medium and long-term forecasting can be achieved using the time series approach with historical power generation data or by converting medium and long-term weather forecast data or reanalysis data. The time series approach requires high-quality historical data. Medium and long-term weather forecast data or reanalysis data require great computational effort, and forecasting accuracy is difficult to guarantee.

Medium and long-term forecasting for offshore wind power generation is mainly used to guide the establishment of regular maintenance plans for offshore wind farms, the designation of monthly and annual power generation plans for power systems, and medium and long-term power market transactions. Due to the limited maintenance time window, poor accessibility, and high operating costs of offshore wind farms, meticulous and coordinated planning and scheduling are required for monthly, quarterly, and annual maintenance and operational calls. As an important support for the operation and management plans, the importance of medium and long-term forecasting for offshore wind power is more prominent. At the same time, with the gradual regulation of the electricity market, medium and long-term power forecasting will become an important measure for offshore wind farms to participate in the medium and long-term market and increase their operating revenue.

5.3.2 Divided by Forecasting Results

5.3.2.1 Deterministic Forecasting

Deterministic forecasting (also called single-point forecasting) is currently the most widely used form of wind power forecasting results, aiming to give a predicted value of wind power at a certain point in the future, which in a mathematical sense can be understood as the expected or median value of future wind power.

According to the modeling principles, forecasting methods can be divided into physical and statistical methods.

(1) The physical method depicts the flow field by establishing an aerodynamic model considering factors such as topography, roughness, altitude, and flow, using NWP as the boundary condition to project the wind speed and direction

at the hub height of the wind turbine, and converting the predicted wind speed into a single point forecasting value according to the theoretical power curve of the wind turbine or the fitted power curve. The advantage of this method is that no historical data is required, and it is suitable for new wind farms; the disadvantage is that the technical threshold of computational fluid dynamics (CFD) is high, and the calculation volume is large. For offshore wind farms, the sea-air-wave coupling effect and the wake effect are obvious, leading to challenges in flow field simulation, which is difficult to meet the accuracy and timeliness requirements of offshore wind power forecasting. To solve the above problem, Li et al. from North China Electric Power University proposed a wind power forecasting method based on CFD pre-calculated flow fields. This method uses numerical simulation and spatial discretization of the wind farm flow field to achieve downscaling of mesoscale numerical weather forecast data. The power forecasting process is converted into a table lookup interpolation process with numerical weather forecast data as input by establishing a database of flow field characteristics. This approach improves the prediction timeliness while ensuring forecasting accuracy.

(2) On the contrary, the statistical method has the advantages of fast calculation speed, high accuracy, strong generalization ability, and a wide range of model selection, making it the most commonly used forecasting method. It establishes the functional relationship between input and output variables by using a large amount of historical data and predicts future wind power based on the trained model. Statistical methods can achieve high prediction accuracy and generalization ability with sufficient historical samples. A review of existing research on wind power forecasting based on statistical methods shows that this field has undergone two development stages.

- The algorithm-focused development stage focuses on using regression algorithms/probabilistic methods with stronger learning abilities or using optimization algorithms to optimize model parameters. For deterministic wind power forecasting, there are two types of commonly used algorithms: neural network algorithms (such as shallow neural networks like Back Propagation, deep learning algorithms such as Deep Neural Networks) and kernel learning algorithms (such as support vector machines, relevance vector machines, extreme learning machines, Gaussian processes).
- The refined modeling-based development stage focuses on improving model adaptability. Many researchers have focused on refined forecasting methods by grouping training samples according to weather conditions or modeling different frequency components of wind speed to enhance the model's adaptability to the "current" wind conditions at a specific time. For wind farms with significant seasonality, refined forecasting models or different forecasting algorithms can be developed directly based on the month/season. For wind farms with complex terrain or a large area, the actual wind conditions are more complex, resulting in insufficient accuracy and representativeness of NWP at a single location, making it difficult to simulate the entire flow

field. In this case, the correlation between wind conditions and the number of NWP points needs to be considered.

5.3.2.2 Uncertainty Forecasting

From a mathematical perspective, the single-point forecast result is the mathematical expectation of future wind power and is the most intuitive expression of future power status. However, single-point forecasting cannot provide information on forecasting uncertainty or risk, and the probability of error is 100%. This limitation greatly restricts the application value of this technology in the wind farm operation and control, new energy power system operation and dispatch, and power market bidding. The value of uncertainty forecasting increases due to the higher risks and challenges associated with offshore wind power forecasting.

Uncertainty forecasting is an extension of single-point forecasting and is used to assess the risk of single-point forecast results, expressed in terms of probabilistic information (probability densities, quantiles, probability intervals), power generation scenarios, and risk indices. Probabilistic information is the most widely used because of its comprehensiveness, intuitive presentation, and visibility. It is mainly used to calculate the range and probability of deviation between the single-point forecast result and the actual power, also known as probability forecasting or interval forecasting.

There are many types of uncertainty analysis methods for wind power forecasting; two commonly used methods are as follows.

(1) Parametric methods

Assume that the error band of the single-point forecast at future moments is a fixed historical empirical value, or assume that the error distribution obeys a specific form of distribution, such as Gaussian distribution, Beta distribution, Versatile distribution, and mixed bias distribution. In practice, the wind power forecasting error does not obey any known functional form. There may be significant differences in the forecasting results of different fitting functions, limiting the universality of the method. Many methods use a segmented fit approach, dividing the data sample into intervals according to wind speed or wind power to improve the fitting accuracy. However, this approach ignores the frequency information in the finer intervals, and the fitting results vary with the segmentation method, making it difficult to describe the actual error objectively.

(2) Non-parametric methods

Non-parametric estimation methods are used to estimate the conditional distribution of error, map learning of error based on artificial intelligence algorithms, or construct uncertainty confidence intervals based on scatter data generated by random simulations. These methods do not require assumptions about the distribution form of the prediction target and can help improve the expression accuracy of the forecasting error distribution. Commonly used methods include quantile regression, adaptive resampling, kernel density estimation, support vector machines, Monte Carlo, sample entropy, cloud model, etc.

Yan et al. from North China Electric Power University proposed a short-term power uncertainty forecasting method for offshore wind farms based on the dynamic cloud model. The dynamic estimation of the forecasting uncertainty is achieved by building a cloud model of the wind power forecasting error at each operating moment and calculating the cloud droplet quantile at any confidence level. After quantifying the forecasting uncertainty, the cloud droplet quantile and the corresponding confidence range for the possible fluctuation of the predicted power at an extreme confidence level are calculated. Combining the inverse and forward cloud generators of the cloud model, the dynamic cloud model is established by selecting training samples in real-time based on the predicted power value at any future moment, and the short-term wind power uncertainty analysis is conducted from both qualitative and quantitative perspectives to improve the reliability of prediction results.

Zhang et al. from North China Electric Power University proposed an improved deep mixed density network for joint uncertainty prediction of regional wind farms. The method introduces a mixed Beta distribution instead of the traditional mixed Gaussian distribution and controls the shape of the Beta distribution to avoid density leakage issues.

5.4 Numerical Weather Prediction for Offshore Wind Power Forecasting

5.4.1 Mode of Offshore Numerical Weather Prediction

Numerical Weather Prediction (NWP) is a mature quantitative forecasting method that uses large-scale computational equipment to solve a set of hydrodynamic and thermodynamic equations describing weather changes and atmospheric motion under certain initial and marginal conditions and to predict the state of atmospheric motion and weather phenomena for a certain period in the future. With the development of recent observation technology and the continuous improvement of computer performance, NWP technology has been developed in leaps and bounds, and prediction accuracy has been continuously improved. As NWP can provide more accurate values of future weather parameters, short- and medium-term wind power forecasting rely heavily on it. Gridded NWP products such as local wind speed, wind direction, air temperature, temperature, humidity, and barometric pressure have been widely used for short- and medium-term wind power forecasting. The development of NWP models for wind power forecasting is mainly based on theoretical research and relevant application technology development of regional models. The hot topics include data assimilation technology, model refinement technology, and physical process parameterization schemes.

According to the different forecast ranges, NWP can be divided into global NWP and mesoscale NWP. The global NWP models operate in about 15 weather service stations, such as Global Forecast System (GFS) and European Centre for Medium-Range Weather Forecasts (ECMWF). The mesoscale NWP models take the output of the global NWP models as input and forecast according to the characteristics of different regions, such as Fifth-generation Mesoscale Model (MM5) and Weather Research and Forecasting Model (WRF).

Currently, the most commonly used regional numerical weather prediction model is WRF. WRF is a new generation mesoscale numerical model and data assimilation system jointly developed by the National Center for Atmospheric Research (NCAR), the National Centers for Environmental Prediction (NCEP), the National Oceanic and Atmospheric Administration (NOAA), multiple universities, research institutes, and operational units. It has the advantages of flexibility, easy maintenance, scalability, and wide applicability, combining advanced data assimilation technology, powerful nesting capabilities, and advanced physics processes, especially in the treatment of convection and mesoscale precipitation. WRF has a wide range of applications for operational numerical weather prediction and atmospheric numerical simulation research, including data assimilation, physical process parameterization, regional climate simulation, air quality simulation, and air–sea coupling simulation.

In order to incorporate the latest scientific research, the WRF modeling system has developed two dynamic frameworks, namely WRF-ARW and WRF-NMM, led by the US NCAR and NCEP, respectively. ARW and NMM are both included in the WRF base software framework. They share the same WRF modeling system framework and physical process modules, except for their different dynamical solver methods. As two independent models in the WRF modeling system, ARW and NMM can be used for real-time numerical weather forecasting and forecasting system research, atmospheric physics and parameterization research, weather phenomenon research, and dynamic framework and numerical weather prediction education. In addition, ARW can also be used for regional climate and seasonal time scale studies, global atmospheric numerical simulation, multi-scale ideal experiments, and data assimilation.

The Global/Regional Assimilation and Prediction System (GRAPES) is a new generation multi-scale general numerical prediction system model jointly developed by the China Meteorological Administration (CMA) and several research institutes. The GRAPES multi-scale unified model can be used for both research and operational purposes and can simultaneously conduct mesoscale/short-term regional/ global medium-term weather forecasting, regional and global climate prediction, simulation and forecasting of dust storm aerosol environment and hydrological flood events, with horizontal scales ranging from 1 to 100 km, and time scales ranging from several hours to two weeks, months, seasons, and years.

Single-atmosphere models have shown good operational application for onshore wind farm forecasting, but when applied to offshore weather forecasting, they do not provide a complete description of the processes between the atmosphere and the ocean. Single-atmosphere models always assume a constant Sea Surface Temperature (SST). However, in practice, the SST tends to vary with atmospheric humidity and

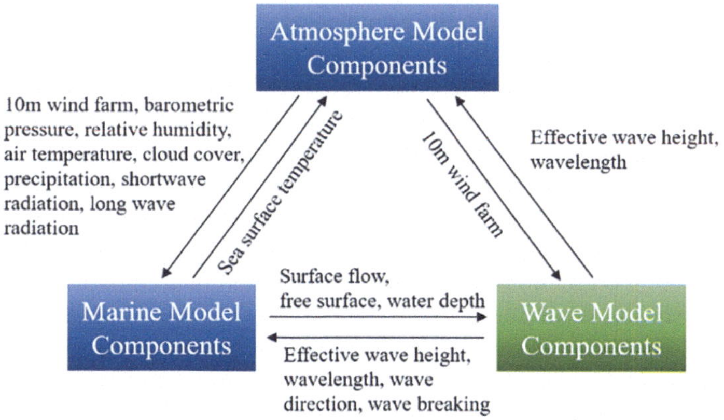

Fig. 5.2 Schematic diagram of coupled mode variable exchange

is also influenced by physical quantities such as shortwave radiation and latent heat fluxes. At the same time, there is a negative feedback mechanism from the SST to the atmosphere, with changing SST affecting precipitation, atmospheric nodal stability, the ocean wind field, and even tropical cyclones to varying degrees. The wave field can further adjust the state of the sea surface, affecting the processes of momentum, heat, and material exchange at the sea-air interface. Therefore, the coupled regional sea-air-wave model is the obvious choice for accurate offshore wind speed forecasting, especially for low-level wind speed at the sea surface.

The regional ocean–atmosphere-wave coupling model takes regional meteorological models, ocean circulation models, and wave forecast models as component models and achieves a large amount of data transmission and exchange among sub-models through a mesoscale coupler. In the model, the atmospheric, oceanic, and wave component models all adopt an open computing framework that allows bi-directional parameter transfer and data exchange with each other, i.e., coupled control functions are added to and called from the integral control functions of their respective component models. At present, a series of coupled models have been proposed by domestic and international scholars for offshore wind field and typhoon forecasting, such as COAMPS, COAWST, GRAPES_TCM-ECOM-WW3, ROMS-WRF-SWAN, IFS-NEMO, as shown in Fig. 5.2.

5.4.2 Artificial Intelligence Correction for NWP

NWP can provide meteorological data such as wind speed and temperature required for wind power forecasting at the same time. The spatial and temporal resolution is usually 3–9 km and 1 h. However, micro-scale effects such as wake, waves, and roughness cannot be reflected in the flow field described by the NWP, which limits the prediction accuracy. The cubic relationship between wind speed and wind power makes small errors in meteorological predicts resulting in large power prediction errors. This accuracy issue is even more pronounced in offshore wind power prediction due to insufficient spatial and temporal resolution. This is mainly because, on the one hand, the tail current effect is prominent in offshore wind farms, and wind power forecasting requires more information about the local flow at the point of the wind turbine; on the other hand, offshore mesoscale forecasting models are more complex, and the contradiction between the timeliness and accuracy of the model calculation is more obvious. Therefore, improving the NWP spatiotemporal resolution and making corrections to the NWP wind speed is necessary to improve the accuracy of offshore wind power forecasting.

Figure 5.3 shows the measured wind speed curve and the NWP wind speed curve at each wind turbine in a certain wind farm. In the figure, the dashed line is the measured wind speed curve at each wind turbine; the thick line is the average measured wind speed curve, indicating the overall trend of measured wind speed; the solid thin line is the NWP wind speed curve at each wind turbine; the thick solid line is the average NWP wind speed curve, indicating the overall trend of NWP wind speed. As can be seen from the graph.

- There is a significant difference between the NWP wind speed and the measured wind speed curve, known as the overall trend deviation, which can reflect the accuracy of basic data provided by the weather prediction department, such as the initial field.
- The measured wind speed curves at each wind turbine location have a large difference, but the corresponding NWP wind speed curves have almost identical changes in trend. When analyzing using the wind speed correlation index at each position, the NWP wind speed correlation is high at each point, while the

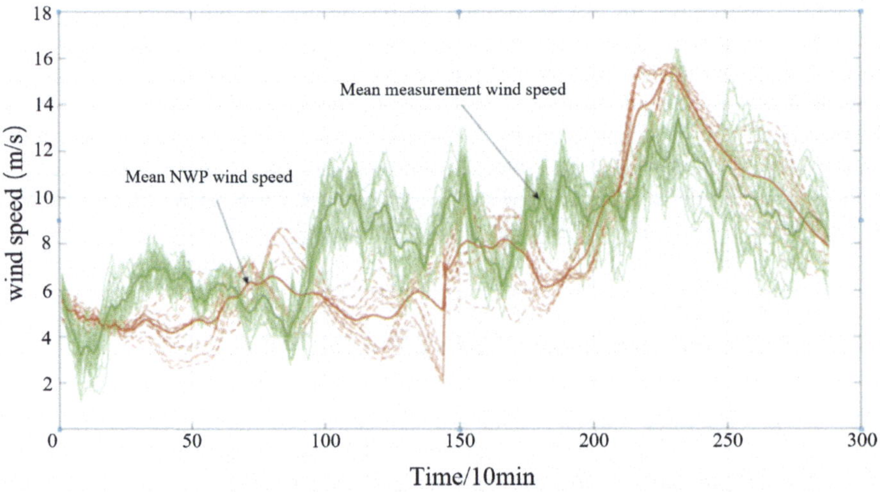

Fig. 5.3 Measured and NWP wind speed time series at each wind turbine in a wind farm

measured wind speed correlation is low and varies among different points. This is the type of error caused by the insufficient spatial resolution of the NWP, known as the spatial distribution deviation.

In summary, in essence, there is a continuous, complex non-linear error between the measured wind speed and NWP wind speed curve caused by various factors, which manifests itself in time as a superposition of overall trend deviations and spatial distribution deviations determined by local characteristics, and in space as spatial distribution errors. The error in NWP has a certain pattern, which is predictable. Therefore, by selecting specific methods to reflect the corresponding differences and fluctuations of NWP at different locations, the NWP error at that location can be corrected.

The technical principle of offshore NWP correction is to build up a large amount of historical wind measurement data and map it to the forecast results to obtain a flow correlation description of the actual flow field, thus improving the accuracy and spatiotemporal resolution of NWP, which can also be understood as a statistical downscaling method. Based on the model input at the time of correction, i.e., the spatial–temporal range of NWP data and the range of correction objects, the study can be divided into NWP correction at a single location, NWP correction considering spatial coupling properties, NWP correction considering temporal relationships, and NWP correction for wind speed and irradiance combined.

(1) **NWP correction at a single location**

NWP wind speed is corrected by establishing a mapping relationship between the NWP data at a specific location (usually a wind measurement tower or a specific wind turbine) and the measured data. This method mainly eliminates the overall trend error of the NWP at independent location points, and it is difficult to consider the correlation between meteorological elements at multiple location points on a spatial scale, thus eliminating the spatial distribution error caused by local effects. This type of error is difficult to eliminate through the correction at a single location.

To improve the correction accuracy, some scholars have used different weather conditions to partition the measured data and NWP data mapping relationship based on wind speed, wind direction, clear sky index, atmospheric stability, and season, or using clustering and scene reduction methods, to establish correction models under different weather conditions.

Some scholars also decompose the measured and NWP wind speed series into different frequency domains from the frequency domain perspective and establish the mapping relationship between measured and NWP data in each frequency domain, then combine the results obtained from each frequency domain to obtain the correction results of NWP wind speed. Commonly used decomposition methods include empirical mode decomposition, variational mode decomposition, wavelet decomposition, etc.

(2) **NWP correction considering spatial coupling properties**

This type of correction considers the wind speed coupling relationships at different spatial locations. When correcting the NWP at a certain point, in addition to the NWP and measured information at that point, NWP information or measured information in neighboring areas is also introduced, i.e., NWP corrections are made by establishing a mapping relationship between NWP data at multiple points and measured data at a single point, or between NWP data at multiple points and measured data at multiple points. Multi-source and multi-point corrections can also be considered by incorporating NWP results from multiple sources.

(3) **NWP correction considering temporal relationships**

The above NWP correction methods are based on the prior statistical law of wind speed prediction error at independent moments, i.e., by establishing a mapping relationship between NWP data and measured data at the same moment in the historical series. In fact, as can be seen from Fig. 5.2, the NWP error patterns at any two moments are very different, and even the measured data corresponding to the same NWP wind speed can vary greatly, resulting in a strong stochastic uncertainty in the mapping relationship between the input and output variables of the correction model. This is one of the important reasons why the existing NWP correction accuracy is difficult to improve.

Therefore, when modeling NWP corrections, it is not enough to establish the mapping relationship with the measured data at the same time, but more influential factors need to be taken into account. In fact, there is a certain transfer relationship within the time series of weather elements, which is also the theoretical basis for the time series method to predict wind speed and power. The Sequence Transfer Correction Method (STCM) for NWP wind speed was proposed by Wang et al. from North China Electric Power University. The input variables of this correction method model include the measured data at time t and the NWP data at time $t + 1$. By introducing the transfer relationship within the wind speed time series, the certainty of the mapping relationship between input and output variables is improved, which improves the accuracy of the NWP correction and is applied to wind farm power forecasting.

(4) **NWP correction for wind speed and irradiance combined**

Being in the same driving field, each weather element has varying but clearly existing spatial and temporal correlations under a given weather system. Using the correlation between wind speed and irradiance, Zhang et al. from North China Electric Power University proposed an NWP correction method that integrates NWP wind speed and irradiance. The input variables of the method include NWP wind speed and irradiance. By introducing a spatial–temporal coupling relationship between wind speed and irradiance at multiple points, the certainty of the mapping relationship between the input and output variables is improved, thus improving the NWP correction accuracy and applying it to the power forecasting of centralized wind farms and solar plants.

5.5 Centralized Power Forecasting Method and System for Offshore Wind Farms

5.5.1 Advantages

China's offshore wind power is characterized by large-scale concentrated development, with numerous wind farms located in the same area. There are unique spatial and temporal correlations between wind conditions and output within the wind power bases. By considering the correlation information of the internal weather conditions and linking meteorological and power generation data, a cluster spatiotemporal joint forecasting model can provide simultaneous forecasting results for all wind farms and wind turbines within the wind power base. This approach can greatly improve the efficiency of forecasting modeling and system maintenance and increase the accuracy of each forecasting unit. Specifically, centralized power forecasting for large-scale offshore wind farms has three advantages:

(1) it can fully consider the strong spatial and temporal correlation of offshore wind power and improve forecasting accuracy;
(2) by consolidating data and forecasting models from multiple wind farms into a unified forecasting system, it can reduce the workload of data processing, model building, forecasting system maintenance and reduce system costs;
(3) the integration of multiple forecasting spatial scales (regional – wind farm – wind turbine) provides convenient and more profitable means for subsequent active grid support control and resource integration to participate in power market bidding.

5.5.2 Principles

In terms of modeling approaches, existing power prediction models typically only use NWP from a single source and location as model inputs to establish a mapping model that corresponds to the output power of a wind farm (or wind turbine). This single-to-single mapping model structure essentially assumes that the NWP at a representative location describes the average wind conditions throughout the entire wind farm and cannot reflect the spatiotemporal flow correlation within the wind farm, ignoring the output correlation between wind turbine units or wind farms, resulting in significant errors in wind power forecasting. This modeling structure can exacerbate the impact of the above errors, especially for compact offshore wind farms.

The current main technical approach for centralized prediction is to use deep learning algorithms that can achieve a multi-to-multi mapping architecture. The joint distribution of NWP from multiple locations represents the wind conditions in the area where the wind farm or multiple adjacent wind farms are located. By relying on the powerful data feature extraction ability of deep neural networks, the extraction of spatiotemporal correlation features among multiple wind farms in the region can be achieved, while in the training process of the prediction model, the output correlation between different wind turbines and different wind farms can be linked to reducing the risk of error mapping, making the power prediction results more reliable and accurate.

Centralized forecasting can achieve predictive modeling at various time scales, NWP correction at multiple points, and joint forecasting of different power types. The main algorithms used include spatial temporal graph convolutional networks (ST-GCN), convolutional long short-term memory neural networks (ConvLSTM), stacked denoising auto-encoders (SDAE), etc. The results show that centralized forecasting methods can accurately explore the spatial–temporal correlations between wind turbines/wind farms in a region and improve forecasting accuracy at the regional level/site level.

It should be noted that the power forecasting model with multi-to-multi mapping has a more stringent sample selection and requires most wind turbines in the wind farm to be in normal power generation state simultaneously. Effective training samples will become very rare if the wind farm is often in a power-limited state or wind turbines frequently fail. The learning ability of deep learning will not be reflected if the training samples are insufficient.

5.5.3 *Centralized Power Forecasting System*

The Wind Farm Laboratory of North China Electric Power University has independently developed a centralized power forecasting system for regional offshore wind farms, which serves all wind farms and all wind turbines within the wind power base and supports various aspects of the wind farm(s) control, power system scheduling, power market bidding, and wind farm equipment operation and maintenance. The system covers real-time forecasting, ultra-short-term forecasting, short-term forecasting, medium and long-term forecasting at wind turbine level, wind farm level, and regional level. It can provide deterministic and uncertainty forecasting, with functions such as ramp events forecasting and automatic model correction.

The main modules of the system include data acquisition, data storage, data cleaning, model training, power forecasting, result correction, and reports and displays, as shown in Fig. 5.4. Each module contains multiple sub-modules to achieve comprehensive data cleaning and forecasting functions.

The topological structure of the system mainly involves a data transfer and download server, reverse isolation device, and group forecasting system server, as shown in Fig. 5.5. The data transfer and download server collects the input data for the forecasting models, including NWP, satellite inversion results, and LiDAR observations, and transmits them to the group forecasting system server via a reverse isolation device. After obtaining the operational data of the wind turbines and wind farms in Safety Zone 1, the forecasting system trains the forecasting model and provides different types of results under different time scales. The forecasting results interact

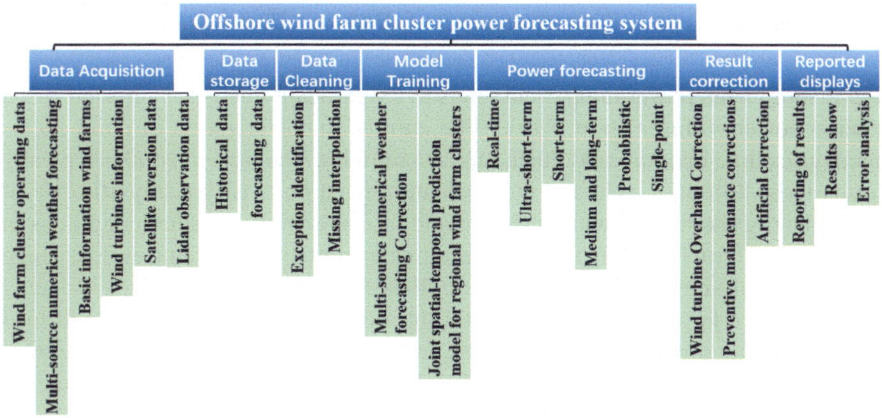

Fig. 5.4 Module components of the centralized power forecasting system for offshore wind farms

with other decision systems; the short-term and ultra-short-term results are reported to the grid agencies.

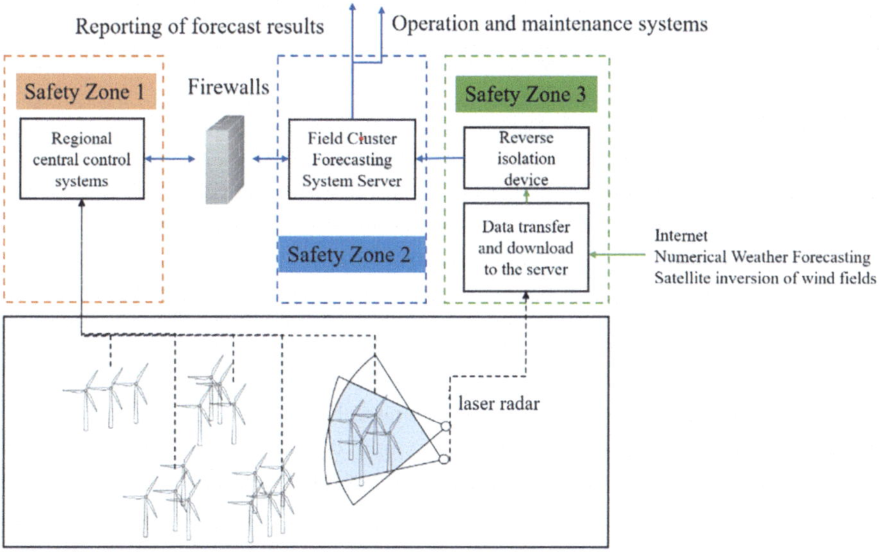

Fig. 5.5 Topology of the centralized power forecasting system for offshore wind farms

5.5.4 Ramp Events Forecasting in Offshore Wind Farms

The offshore wind power ramp event refers to a significant fluctuation in wind speed or wind power within a short period. The current characterization of ramp events mainly uses time intervals and fluctuation magnitudes. The time interval is typically chosen to be 30 min to 6 h, and the fluctuation magnitude is typically 10–75% of rated capacity. Compared with onshore wind farms, extreme weather conditions in offshore wind farms are more frequent, resulting in more frequent ramp events, which can significantly impact the power system. Especially when strong winds suddenly pass through, a large number of wind turbines may suddenly drop from full load operation to shut down, posing a huge risk to the power system. Therefore, to ensure the safe and stable operation of the grid and to develop timely response strategies, accurate prediction of ramp events is essential for offshore wind power.

Existing research has shown that the characteristics of wind power ramp events are influenced by a combination of meteorological conditions in a complex way, such as subsurface, time of year, and season. Some scholars have studied the dependence between offshore wind power ramp events and different weather patterns and estimated whether future ramp events will be generated combined with NWP. However, this method depends on the division way of weather scenarios and the criteria for determining ramp events occurrence, resulting in poor robustness. Other scholars have established mapping relationships between historical weather data, NWP data, and feature parameters of historical ramp events data through statistical learning methods such as neural networks to achieve accurate ramp prediction.

Sun et al. from North China Electric Power University proposed a wind power ramp events prediction method based on the orthogonal experiment and support vector machine, which concluded that the occurrence characteristics of wind power ramp events are influenced by various factors such as meteorological conditions, geographical location, topographic conditions, time of day, and season. Moreover, the influencing factors and their impact on ramp events are different in different regions and wind farms. The inputs of the prediction model can directly affect the prediction accuracy. Therefore, it references the idea of orthogonal experiment design and analysis and introduces the input selection link—orthogonal test in support vector machine, through the orthogonal test to quantitatively analyze the order of the influencing factors in wind farms and optimize the input quantity of the prediction model to achieve high-precision prediction of wind power ramp events.

Chapter 6
Smart Operation Control Technologies for Offshore Wind Farms

6.1 Introduction

Smart operation control for offshore wind farms is the core technology to improve the total generation capacity, reduce the fatigue load of wind turbines, enhance the friendliness and support capacity to the grid, and maximize the profitability of the whole life cycle of wind power.

This chapter introduces three key objectives and cutting-edge directions of smart operation control technologies for offshore wind farms, including wake control for power enhancement, fatigue load control for wind turbine life extension, and optimized operation control for ancillary grid services.

6.2 Challenges of Operation Control for Offshore Wind Farms

Compared with onshore wind farms, the wake effect of offshore wind farms is more pronounced and complex, leading to up to 40–50% higher loss of entire power generation. This is mainly due to the following reasons: The smoother sea surface and lower sea surface roughness of offshore wind farms lead to smaller vertical wind shear, resulting in lower environmental turbulence and longer wake recovery distance; The relatively stable atmospheric environment at sea causes slower interaction of turbulence flux in the vertical direction, weakening the rate of wake recovery; Furthermore, the investment and operating costs for transmission lines in offshore wind farms are significantly higher than those for onshore wind farms, so most offshore wind farms have a centralized and regular layout of wind turbines, which further exacerbates the wake effect.

Offshore wind turbines are subjected to various loads, including wind, waves, currents, and ice, due to the influence of natural environmental factors such as tides,

© The Author(s), under exclusive license to Springer Nature Singapore Pte Ltd. 2025
Y. Liu et al., *Offshore Wind Farm Technology*,
https://doi.org/10.1007/978-981-96-1889-7_6

waves, currents, vortices, floating ice, and typhoons. They also have larger capacities and longer blades compared to onshore wind turbines, which results in a greater load-bearing capacity requirement in high wind speeds and harsh offshore environments. This dramatically affects the availability of offshore wind turbines and the profitability of offshore wind farms.

China's offshore wind power is mostly connected to the eastern power grid. Due to the large-scale feed-in of ultra-high voltage transmission in the west, the local synchronous power supply is significantly reduced, and the system backup capacity is insufficient. Thus, large-scale offshore wind power connected to the grid has a greater impact on the safe and stable operation of the power system.

6.3 Wake Control of Offshore Wind Farms for Power Enhancement

The wake effect refers to the downstream region where wind speed decreases after wind turbines extract energy from the wind. If there is a downstream wind turbine located in the wake zone, the input wind speed of the wind turbine is lower than that of the upstream wind turbine. If the upstream wind turbine uses a traditional single-unit maximum power tracking control strategy, the downstream wind turbines in the wake zone will be adversely affected in terms of inflow wind conditions, operating status, and power generation.

The wake control technology for offshore wind farms can reduce the power loss caused by wake effects, significantly improve the overall power generation capacity, and is a cutting-edge technology in the field of offshore wind farms. It is also a key supported area for research programs in countries such as China, Norway, Denmark, the United States, and the European Union (EU). In 2018, the EU set up the Total Control Project with a budget of 480 million euros. In 2019, China and Norway launched an inter-governmental cooperation project of "Research on Smart Operation Control Technology for Offshore Wind Farms". The research on wake control in offshore wind farms can be mainly divided into fast wake calculation models, integrated models, control models, and fast optimization algorithms.

6.3.1 Fast Wake Calculation Models

Accurately simulating the wake distribution of wind turbines within a wind farm is the basis of wake control and directly determines the actual effect of wake control. Although the traditional Computational Fluid Dynamics (CFD) method has high calculation accuracy, it is time-consuming and inefficient, making it difficult to meet the timeliness requirements of control strategies. Therefore, there is an urgent need to research high-precision fast wake calculation methods, including engineering models, CFD pre-calculated methods, and multi-precision simulation techniques. Currently, research and engineering practices at home and abroad mainly focus on engineering wake models. The inflow wind conditions for wake calculation can use the measured wind speed or the real-time wind speed forecasting results.

6.3.1.1 Engineering Wake Models

(1) Engineering wake models for single wind turbine

The engineering wake model used for wind farm wake control is a low-precision wake distribution calculation model that combines the engineering wake model (analytical wake model, empirical wake model) for a single wind turbine with the engineering superposition model [1]. It has a simple form and requires fewer computational resources, but the calculation accuracy depends on the adjustment of empirical parameters in different operating conditions. The model focuses on calculating velocity distribution in the far wake region of the wind turbine and has a lower accuracy in calculating velocity in the near wake zone.

Currently, the commonly used engineering wake models for the single wind turbine are the Jensen wake model based on mass conservation, the Frandsen wake model based on momentum theorem, and the Larsen wake model obtained by solving simplified Prandtl turbulent boundary layer equations. Katic [2] supplemented the Jensen wake model based on the one-dimensional momentum theory and derived the widely used Park wake model in engineering. After years of engineering practice verification and improvement, the calculation results can already meet the engineering requirements. Bastankhah and Porté-Agel used the theorem of momentum and Gaussian distribution function to study the two-dimensional form of the Frandsen wake model (BPA model), which has been widely used in academic research.

(2) Engineering wake models for single wind turbine considering yawing

Yawing action is required in wake control, where the wake centerline will shift when the wind turbine is yawing, thereby changing the range of wake impact and affecting the output of downstream wind turbines. In such a scenario, a wake deflection model is needed to calculate the displacement of the wake centerline. Based on this, the engineering wake model for a single wind turbine can be modified to more accurately simulate the velocity distribution in the wake zone when the wind turbine is yawing. The commonly used wake deflection models include the Jiménez model, the Bastankhah model, the Shapiro model, etc. [3].

(3) Wake superposition models

Offshore wind farms are often composed of dozens or even hundreds of wind turbines arranged in a concentrated layout, and downstream wind turbines may be simultaneously affected by the wake superposition of multiple upstream wind turbines. The flow characteristics in the wake superposition zone are extremely complex and involve complex turbulence mixing mechanisms. The wake effect is not simply the sum of the wake effects from each wind turbine.

The most commonly used approach in engineering is to combine the single wind turbine engineering wake model with a superposition model to simulate the wake distribution in a wind farm. There are the four commonly used superposition models: geometric sum (GS), linear sum (LS), energy balance (EB), and quadratic sum (QS). The QS was proposed by Katic in a classic literature about the Park wake model, assuming that the velocity loss in the wake superposition region is equal to the square root of the sum of the squared velocity losses in the wake regions of each upstream wind turbine. Currently, the QS combined with the classic Jensen wake model is a commonly used low-precision wake distribution calculation method. Shao et al. from North China Electric Power University proposed a k-based superposition model and a modified energy balance (MEB) superposition model to address the issues with the QS superposition model. They combined these models with the Jensen model and verified them using power measurements from Lillgrund and Horns Rev I offshore wind farms. Compared with other mainstream models, the models could more accurately simulate changes in wind turbine power output.

In summary, in current academic research and engineering practice, few higher-precision wake superposition models can be applied to engineering, which will still be a focus and challenge in future research.

6.3.1.2 CFD Pre-calculated Flow Fields Method

The CFD method based on numerical discretization to solve the Navier–Stokes equations is the most accurate method for simulating the aerodynamic characteristics of the wake region. It not only provides detailed information about the velocity, pressure, and density distribution in the wake region of wind turbines but also accurately simulates the influence of complex turbulent flow on the wake characteristics. However, the CFD method requires an enormous computational cost to simulate the wake field of wind turbines, and its computational efficiency is far from meeting the requirements of wake control.

To overcome the contradiction between the calculation efficiency and accuracy of the CFD method in simulating wind turbine wake fields, Li et al. from North China Electric Power University proposed a concept of CFD Pre-calculated Flow Fields (CPFF) [4]. First, the method involves offline CFD modeling to calculate the wake region characteristics under different inflow and operating conditions and then generate a database. Then, based on the database, the wake region distribution information under any target operating conditions can be quickly obtained by combining proxy algorithms, such as statistical interpolation and artificial intelligence. This method has been successfully applied in wind speed prediction [4], wind power prediction [5], wake modeling, wake control [3], etc., and can effectively balance the contradiction between calculation efficiency and accuracy. The selection of representative wind and operating conditions for database construction and the reduction of computational complexity required to construct the database while ensuring calculation accuracy are important directions for the future development of this technology.

6.3.1.3 Multi-precision Wake Simulation Technology

The multi-precision wake simulation technology has emerged to achieve a balance between computational accuracy and efficiency in the simulation of wind turbine wake regions. Based on CPFF, this technology significantly reduces the time required to obtain wake field information compared to using high-precision CFD methods alone, allowing for fast wake simulation under any operating conditions. However, the time cost of building a high-precision CFD method to obtain wake field information databases is still high, which to some extent, limits the size of the CFD wake pre-calculation database and affects the final wake field calculation accuracy and corresponding control effects. To achieve a balance between computational accuracy and efficiency in the simulation of wind turbine wake fields, it is necessary to further reduce the number of CFD simulations under all wind conditions, thus giving development to the multi-precision wake simulation technology.

Wang et al. from North China Electric Power University proposed a multi-fidelity (MF) approach for simulating wind turbine wake fields. This approach combines an engineering wake model with lower precision but faster calculation and CFD calculation results that can contain more wake field details and higher calculation accuracy to obtain a high-level pre-calculated model with more balanced calculation accuracy and efficiency.

Compared to single-precision models using the same number of high-precision sample points, the MF wake flow simulation technique can use fewer high-precision sample points to obtain higher computational accuracy, making it easier to capture changes in the target function trend. It can significantly reduce the number of samples required for high-precision wake flow models and the time required for calculations under all operating conditions. This provides more possibilities for the application of time-consuming CFD models in wind farm wake control and other fields. However, similar to the pre-calculated method, selecting representative high-precision sample points and further balancing the relationship between accuracy and computation is an important direction for the future development of this technology.

6.3.2 Integrated Model

The wake control integrated system of offshore wind farms is a fully integrated and developed simulation tool that includes real-time wind speed forecasting technology for wake control, electrical modules of wind turbines, fast wake calculation modules, wake optimization control algorithms, data post-processing, and visualization functions. It realizes the mutual conversion between wind speed, wake, and power, providing a complete model, powerful function, and friendly secondary development tool for wake control in offshore wind farms. For example, the Floris software toolkit, a low-cost and control-oriented wind farm steady-state wake characteristic modeling tool developed jointly by the National Renewable Energy Laboratory and the Delft University of Technology, is suitable for the design and optimization of wind farm control strategies through the simulation of wake effects in operating wind farms.

The integrated model expresses the process of wind energy absorption by wind turbines, calculates the power of wind turbines based on the power coefficient, and calculates the wind speed distribution in the wake region behind the wind turbine based on the thrust coefficient. There are mainly three ways:

(1) **Theoretical Derivation**

According to the blade element momentum theory and the actuator disk theory, the theoretical formula is as follows.

Axial induction factor a:

$$a = 1 - \frac{u_0}{u_\infty} \tag{6.1}$$

Thrust coefficient C_T:

$$C_T = 4a(1 - a) \tag{6.2}$$

Wind turbine power coefficient C_P:

$$C_P = 4a(1 - a)^2 \tag{6.3}$$

Power P:

$$P = \frac{1}{2} C_P \rho A u_i^3 \cos^3 \gamma \tag{6.4}$$

where, A is the swept area of the wind turbine; ρ is the air density; u_i is the wind speed; γ is the yaw angle of the wind turbine.

(2) **Empirical formula**

The thrust coefficient calculation method is modified when a is greater than 0.5, and the wind turbine operates in the state of disturbance wake. For example, the empirical formula applied in GH Bladed is as follows:

$$C_T = 0.6 + 0.61\mathrm{a} + 0.79\mathrm{a}^2 \tag{6.5}$$

(3) **Measured curves from wind turbine supplier**

When researching a specific wind turbine, C_T-wind speed and C_P-wind speed curves provided by the wind turbine supplier are often used as inputs for the integrated model, which is one of the most widely used approaches currently.

6.3.3 Control Model

To improve the power generation efficiency of an individual wind turbine, the maximum power tracking control strategy is adopted before reaching the rated wind speed. However, in some cases, this strategy can cause downstream wind turbines to suffer greater wake effects, resulting in higher power losses for the entire wind farm. The essence of wake control is to sacrifice the power generation efficiency of upstream wind turbines to reduce wake losses for downstream wind turbines, thereby increasing the overall power generation of the wind farm. Two optimization strategies are mainly adopted: yaw optimization and speed optimization.

Yaw optimization [6] is mainly used to achieve the deflection of the wake region by changing the yaw angle of the upstream wind turbine to reduce the wake effect, which offers the advantages of simple control and high efficiency. It has received extensive attention and in-depth research from enterprises and research institutions, but active yaw will increase the fatigue load of the wind turbine, which may affect the equipment's lifespan.

Speed optimization is mainly achieved by changing the rotational speed of the upstream wind turbine to reduce its power generation and increase the wind speed in front of the downstream wind turbine. In theory, this method causes minimal load damage to the wind turbine but requires a larger degree of modification to the original control strategy of the wind turbine. Related research and engineering applications are rare.

(1) **Yaw optimization**

Yaw optimization is achieved by actively changing the wind turbine's yaw angle to deviate the rotor's centerline from the incoming wind direction, thereby adjusting the wake direction and shifting the wake center away from downstream wind turbines to reduce the wake effect and increasing power generation. The mathematical expression of the wind farm's total power optimization model based on yaw optimization is as follows:

$$P = \sum_{i=1}^{n} P_i = \sum_{i=1}^{n} \frac{1}{2} C_{Pi} \rho A_i u_i^3 \cos^3 \gamma_i \tag{6.6}$$

where, n is the number of wind turbines; P_i is the power of i th wind turbine; C_{Pi} is the power coefficient of i th wind turbine; A_i is the swept area of i th wind turbine; u_i is the wind speed of i th wind turbine; γ_i is the yaw angle of i th wind turbine.

(2) **Speed optimization**

Currently, wind turbines below the rated wind speed mostly use constant tip speed ratio control to ensure maximum wind energy utilization. Changing the rotational speed can adjust the tip speed ratio to decrease wind energy utilization and increase the wind speed in the wake region, coordinating wind turbines' output power and wake loss, thereby maximizing the wind farm's overall power output. Since no universal empirical or theoretical formula directly corresponds to the relationship between speed optimization and wind energy utilization coefficient, the axial induction factor is mainly used as the optimization variable when implementing speed control. The mathematical expression is as follows:

$$P = \sum_{i=1}^{n} P_i = \sum_{i=1}^{n} \frac{1}{2} C_{Pi} \rho A_i u_i^3 = \sum_{i=1}^{n} 2a_i(1 - a_i)^2 \rho A_i u_i^3 \tag{6.7}$$

6.3.4 Fast Optimization Algorithms

Wind farm wake optimization control has characteristics such as high dimensionality, nonlinearity, and coupling of multiple parameters. Due to the complex and variable flow field in wind farms, traditional control algorithms find it challenging to achieve effective wake optimization. Given the complexity of actual wind farm wake control problems, seeking efficient and fast optimization algorithms has become one of the main research topics in the field of wind farm wake control. Mainstream fast optimization algorithms include numerical optimization algorithms, intelligent optimization algorithms, and reinforcement learning algorithms.

(1) **Numerical optimization algorithm**

Numerical optimization algorithms convert the wind farm wake optimization control problem into a numerical optimization problem to maximize the overall output of the wind farm by adding necessary constraints. The best wind farm wake optimization control solution can be obtained by optimizing the objective function. Numerical optimization algorithms have the advantages of simple implementation and fast convergence speed, but challenges exist, such as difficulty finding the optimal global solution.

(2) **Intelligent Optimization Algorithms**

Intelligent optimization algorithms are inspired by human intelligence, biological populations, or natural laws and use intelligent optimization methods to simulate natural phenomena or intelligent behaviors of biological populations. Although individuals in intelligent optimization algorithms only follow simple rules and there is no group center control, the interaction between groups triggers the emergence of intelligence at the global level. Due to their simplicity, efficiency, generality, and strong applicability, intelligent optimization algorithms have been widely used in wind farm wake control. For example, Gu et al. from North China Electric Power University proposed a multi-agent optimal scheduling method for the wind farm, and Shao et al. proposed a wake optimization control method for the wind farm based on particle swarm optimization.

(3) **Reinforcement Learning Algorithms**

Reinforcement learning is a type of machine learning algorithm that adapts to the environment based on feedback. The intelligent agent is in the environment and affects the environment through actions. The environment gives the intelligent agent a certain reward based on the reward function. The intelligent agent continuously interacts with the environment during training, observes the impact on the environment and the rewards obtained, and guides the intelligent agent's future actions. Compared with numerical optimization algorithms and intelligent optimization algorithms, reinforcement learning algorithms have the advantages of self-learning and online learning, making them suitable for solving complex wind farm wake optimization control problems with limited prior knowledge. Wang et al. from North China Electric Power University implemented wind farm wake control based on a deep reinforcement learning algorithm, using control parameters such as yaw angle, pitch angle, and rotational speed of wind turbines as the agent's actions, and power and fatigue load as reference values for the reward function.

6.4 Fatigue Load Control for Wind Turbine Life Extension

Wind turbines are subjected to long-term alternating fatigue loads during operation. Fatigue is major reason for wind turbine failures, reduced service life, and reduced reliability. Offshore wind turbines are subjected to complex coupled loads, including wind, wave, flow, and ice loads, which result in significantly higher fatigue load than onshore wind turbines [7]. Therefore, fatigue load control for offshore wind farms aimed at extending the life of wind turbines can reduce load-reducing actions such as start/stop or pitch adjustment, prolong the service life of wind turbines, and reduce the levelized cost of electricity for the entire lifecycle of the wind farm.

6.4.1 Fatigue Load of Offshore Wind Turbine

6.4.1.1 Wind Load

According to aerodynamic principles, wind generates along-wind force, cross-wind force, and torque due to the obstruction of the structure. For conventional structures, wind mainly acts on the structure as along-wind force. The mathematical expression of the standard value of along-wind force working on offshore wind turbines is as follows [8].

$$W_z = \beta_z \mu_s \mu_z W_0 \tag{6.8}$$

where, W_z is the standard wind load value acting on the unit projected area at height z of the structure, kPa; β_z is the wind-induced vibration coefficient at height z; μ_s is the shape coefficient; μ_z is the wind pressure variation coefficient at height z; W_0

is the basic wind pressure, kPa. The coefficients β_z, μ_s, μ_z can be referred to in the "Design Specification for High-rise Structures".

The calculation formula for the basic wind pressure W_0 is:

$$W_0 = \frac{1}{1600}v^2 \tag{6.9}$$

where, v is the 10-min average wind speed at the height of 10 m above the open ground near the wind turbine, with a 50-year recurrence period.

6.4.1.2 Wave Load

For small-scale piles with D/L or b/L $\leq$ 0.2, when H/d $\leq$ 0.2 and d/L $\geq$ 0.2, or H/d > 0.2 and d/L $\geq$ 0.35, the positive force acting on the full section of the pile above the water surface at height z and parallel to the wave direction is composed of velocity and inertia forces:

$$p_D = \frac{1}{2}\frac{y}{g}C_D D_u|u| \tag{6.10}$$

$$p_1 = \frac{y}{g}C_M A\frac{\partial u}{\partial t} \tag{6.11}$$

where, p_D is the velocity force of the wave force, kN/m; p_1 is the inertia force of the wave force, kN/m; γ is the weight density of seawater, kN/m3, which can be taken as 10.2 kN/m3; g is the gravity acceleration, m/s2, which can be taken as 9.8 m/s2; D is the diameter of the column, m. When the cross-section is rectangular, D is replaced with b; A is the cross-sectional area of the column, m2; u is the horizontal velocity of water particle, m/s; $\partial u / \partial t$ is the horizontal acceleration of water particle, m/s2; C_D and C_M are the velocity force coefficient and inertia force coefficient, respectively, which can be determined by referring to "Hydrplogy Standard Specifications for Port and Harbour" (JTS145-2-2013) and the "Technical Specifications for Beach and Sea Environment Conditions and Load" (SY/T4084-2010).

6.4.1.3 Flow Load

Flow load is an external load caused by the large-scale water flow, mainly in tidal flow, which directly acts on the foundation structure of offshore wind turbines as an ocean environmental load. For offshore wind turbines, the flow load cannot be ignored. The ocean current velocity changes slowly over time. The mathematical expression of the flow load is:

$$f = \frac{1}{2}\rho C_D A u_0^2 \tag{6.12}$$

where, ρ is the density of seawater, which can be taken as 1025 kg/m^3 when there is no measured data; A is the equivalent area of the structure, m^2; u_0 is the flow velocity, m/s; C_D is the drag coefficient.

6.4.1.4 Ice Load

The effect of ice load should be considered for offshore wind turbines in icing areas. According to the interaction between ice and wind turbines, the main forms of ice load are static ice pressure generated when a large area of ice moves under the action of wind and flow and the impact load of the free drifting ice.

(1) Fixed ice load

The mathematical expression for the standard value of the ultimate compressive ice force produced by a large area ice load on the pile or pier is as follows:

$$F_I = ImkBH\sigma_\varepsilon \tag{6.13}$$

where, F_I is the standard value of the ultimate compressive ice force, kN; I is the local compression coefficient of ice; m is the shape coefficient of the pile (pier) facing the ice; k is the contact condition coefficient between the ice and the pile (pier), which can be taken as 0.32; B is the projected width of the pile (pier) facing the ice, m; H is the calculated thickness of a single layer of level ice, m; σ_ε is the standard uniaxial compressive strength of the ice, kPa. The values of these parameters can refer to the relevant provisions in the "Load Code for Harbour Engineering" (JTS144-1-2010).

(2) Drift ice load

The expression for the standard value of the impact force of drift ice on upright cylinders and upright piers is:

$$F_z = 2.22HV\sqrt{IkA\sigma_c} \tag{6.14}$$

where, F_z is the standard value of the impact force produced by the drift ice on the cylinder (pier); H is the calculated ice thickness of the single layer of flat ice; V is the velocity of the drift ice; A is the plane area of the drift ice block; σ_c is the standard value of the uniaxial compressive strength of ice.

6.4.2 Fatigue Damage Calculation Method Suitable for Fatigue Load Control

6.4.2.1 Equivalent Fatigue Load

The equivalent fatigue load spectrum is a random repetitive variation load spectrum converted into a constant amplitude load spectrum. The effect of the equivalent fatigue load is the same as the effect of the accumulated load during the lifetime. The mathematical expression of the equivalent fatigue load S_{equ} is:

$$S_{equ} = \sqrt[m]{\frac{\sum n_i S_i^m}{Tf}} \tag{6.15}$$

where, S_i is the stress range of the rainfall counting cycle; n_i is the number of cycles of the load range S_i; m is the inverse slope in the S–N curve equation; T is the duration of the original time distribution; f is the frequency of the load signal.

6.4.2.2 Sequential Fatigue Damage

The Palmgren–Miner linear damage accumulation theory suggests that the fatigue damage produced by materials under the action of load time series are independent of each other, which do not interfere with each other and can be linearly superimposed. When the accumulated damage reaches a threshold value, material damage occurs. If the structure acts n_i times under k typical load cycles with stress variation $\Delta\sigma_i$, the mathematical expression for sequential fatigue damage is:

$$D = \sum_{i=1}^{k} D_i = \sum_{i=1}^{k} \frac{n_i}{N_i} \tag{6.16}$$

where, D_i is the fatigue damage caused by ith stress variation on the bolt; n_i is the cumulative number of cycles corresponding to the stress variation; N_i is the permissible number of cycles corresponding to the stress variation, which is determined by the S–N curve of the structure; k is the total number of stress variation.

6.4.3 Application of Fatigue Load Optimization Control

The operation and maintenance costs of offshore wind farms are high, and the wake effect is prominent. In addition to affecting the efficiency of power generation, but also increases the fatigue damage of wind turbines, leading to improved operation and maintenance costs. The mainstream method for reducing the fatigue damage of wind turbines is to operate each wind turbine according to the overall output strategy during the free power generation stage of the offshore wind farm and to optimize the fatigue distribution of each wind turbine in the stage of power limitation at the dispatch center, with the minimum fatigue damage or fatigue imbalance as the optimization target for fatigue load optimization control.

6.4.3.1 Optimal Control with Minimum Overall Fatigue Damage

(1) Objective function

Unlike traditional energy generation stations, wind farms do not have coal consumption curves of thermal power plant units or flow regulation of hydroelectric power stations. The power generation of wind farms mainly depends on wind power forecasting. The combination of wind turbines in a wind farm depends on the wind speed conditions at the location of the wind turbine and the grid's ability to accommodate wind power. According to the wind power forecasting results and the grid scheduling instructions, the optimization objective function is to find a combination scheme that can reduce redundant operations and avoid frequent start/stop of wind turbines with the minimum fatigue damage of key components. The objective function is as follows:

$$F = \min\left(\sum_{j=1}^{T}\sum_{i=1}^{N}(a_i^j u_i^j \cdot t) + \sum_{j=1}^{T}\sum_{i=1}^{N} b_i^j u_i^j (1 - u_i^{j-1}) + \sum_{j=1}^{T}\sum_{i=1}^{N} c_i^j u_i^{j-1}(1 - u_i^j) + \sum_{j=1}^{T}\sum_{i=1}^{N}(d_i^j (1 - u_i^j)\cdot t)\right) \tag{6.17}$$

where, F is the total damage; T is the number of periods; N is the number of wind turbines; a_i^j is the fatigue damage when wind turbine i is in normal operation at time j; b_i^j is the fatigue damage when wind turbine i is starting at time j; c_i^j is the fatigue damage when wind turbine i is shutting at time j; d_i^j is the fatigue damage when wind turbine i is idling at time j; u_i^j is the start/stop state of wind turbine i at time j. 0 represents stopping, and 1 represents running; t is the operation time of the wind turbine.

(2) Constraints

The constraints of wind turbines combination are as follows.

1) *Upper and lower limits of wind turbine output*

$$P_{i,\min}^j \le P_{i,predict}^j \le P_{i,\max}^j \tag{6.18}$$

Table 6.1 Change rates of wind farm power

Installed capacity of wind farms (MW)	Maximum variation in 10 min (MW)	Maximum variable in 1 min (MW)
< 30	20	6
30–150	Installed capacity/1.5	Installed capacity/5
> 150	100	30

where, $P^j_{i,predict}$ is the power forecasting value of the ith wind turbine; $P^j_{i,\min}, P^j_{i,\max}$ is the minimum and maximum output of the ith wind turbine.

2) *Load dispatch*

$$\sum_{i=1}^{N}(P^j_{i,predict} \cdot u^j_i) - P_{loss} = P^j_{load} \tag{6.19}$$

where, P^j_{load} is the planned load of the wind farm at time j, meeting the load dispatch command in the power grid.

3) *Maximum change rate of wind power*

The change rates of wind farm power during grid connection, the normal shutdown of wind turbines, and wind speed increase are shown in Table 6.1.

6.4.3.2 Optimal Control with Maximum Uniformity of Overall Fatigue

(1) Objective function

The integrated optimization strategy for fatigue and active power of offshore wind farms is designed as follows:

$$\min(F^{farm}_{st}(t_f)) = \min(\sqrt{\frac{1}{n_{farm}}\sum_{t=t_0}^{t_f}\sum_{i,j}(F^{i,j}(t) - \overline{F^{farm}(t)})^2}) \tag{6.20}$$

where, $F^{farm}_{st}(t_f)$ is the standard deviation of the fatigue coefficient of each wind turbine at the time t_f; $\overline{F^{farm}(t)}$ is the average value of the fatigue coefficient of each wind turbine at the time t; n_{farm} is the number of wind turbines; t_0 is the current time.

(2) Constraints

The active power constraint of the wind farm indicates that the active power should not deviate too much from the reference value, and from the perspective of security and stability of the power system, the smaller deviation is better.

$$\left| P_{ref}^{farm}(t) - \sum_{i,j} P^{i,j}(t) \right| \leq \varepsilon_{farm} \tag{6.21}$$

where, $P_{ref}^{farm}(t)$ is the reference power issued to the wind farm by the grid dispatch agency at time t; $P^{i,j}(t)$ is the active power of the wind turbine $A(i,j)$ at the time t; ε_{farm} is the upper error limit in tracking the reference power of the wind farm.

The wind turbine output constraint represents the active power of each wind turbine should be able to produce as specified by the control system with enforceability.

$$P_{low}^{i,j} \leq P^{i,j}(t) \leq P_{rate}^{i,j} \tag{6.22}$$

where, $P_{low}^{i,j}$, $P_{rate}^{i,j}$ are the minimum power and rated power, respectively.

6.5 Optimized Operation Control for Ancillary Grid Services

The offshore wind farm greatly impacts the stability of the eastern power grid in China. An optimal operation control strategy for an offshore wind farm(s) oriented to auxiliary services of the power system is needed, mainly including system frequency regulation, peak load regulation, and system reserve capacity.

(1) With the development of offshore wind farms towards deep and distant seas, the number of wind turbines connected to the grid by flexible DC is increasing. However, due to the short inertia time constant of wind turbines and the decoupling effect of flexible DC transmission systems, the system inertia of the AC system decreases, and offshore wind farms cannot directly respond to frequency changes in the onshore AC system. To solve the above problems, current research mainly uses rotor kinetic energy control, load shedding control, and cluster control to participate in frequency support.

- Rotor kinetic energy control is mainly through additional inertia control and droop control. When the frequency drops, the kinetic energy stored in the rotor is released by changing the rotational speed to participate in system inertial support. The response is fast. However, the supporting time and capacity are limited in the absence of power reserve.
- Load shedding control is mainly achieved through two methods. The first method is to shift the operating point of wind turbines from the maximum power tracking point to the load shedding point, thereby achieving overspeed load shedding control. This method has a fast response speed. However, similar to rotor kinetic energy control, it still suffers from the problem of small adjustable capacity. Another method is to change the pitch angle of the wind

turbine to achieve variable pitch load shedding control for power reserve. This method has a wide adjustable range and large adjustable capacity. However, it requires adjustments to the mechanical structure, and the adjustment speed may cause damage to the mechanical structure.
- Cluster control is a strategy for a large-scale wind farm(s) with a significant difference in operating conditions. After integrating the system dispatch instructions and wind turbine operating status information received by the wind farm, and according to the different operational characteristics of each wind turbine, a certain allocation strategy is followed to allocate the set value of the frequency regulation active power to each wind turbine. This maximizes the frequency regulation capabilities of the wind farm. The control strategy requires high communication reliability and the selection of iterative optimization objectives.

(2) The volatility and anti-peak characteristics of wind power increase the difficulty of grid peak regulation. Compared with onshore wind turbines, offshore wind turbines are more significantly affected if they are frequently started/stopped due to the characteristics of the operating environment. Therefore, the anti-peak characteristics are more prominent. Under the premise of meeting peak regulation demand, maximizing the power generation capacity of wind farms, and reducing the number of start/stop cycles of wind turbines are the main research contents of offshore wind turbines participating in peak regulation. To solve the above problems, the current research mainly focuses on two aspects.

- Fully utilizing the grid's consumption capacity by optimizing the active power control strategy of the wind turbine.
- Achieving economic operation of the power grid through the coordinated operation of multiple energy sources such as photovoltaic, nuclear power, and hydropower to promote wind power consumption.

(3) Offshore wind power is difficult to be used directly for system backup due to its randomness and volatility, so it needs to be equipped with energy storage. Compared with onshore wind power, on the one hand, offshore weather conditions are harsher; the performance requirements for energy storage equipment are higher in terms of high-voltage resistance, wind and wave resistance, and seawater corrosion resistance. On the other hand, the use of water is more convenient and direct in offshore wind farms. Pumping energy storage, underwater compressed air, and hydrogen storage are more convenient and economical.

References

1. Archer, C. L., Vasel-Be-Hagh, A., Yan, C., et al. (2018). Review and Evaluation of Wake Loss Models for Wind Energy Applications. Applied Energy, 266:1187–1207.

2. Katic, I., Højstrup, J., & Jensen, N. O. (1987). A Simple Model for Cluster Efficiency. In:W. Palz, & E. Sesto (Eds.), EWEC'86. Proceedings , Vol. 1. A. Raguzzi ,Italy ,pp. 407–410.
3. Shao, Z. Z. (2019). Research on Rapid Simulation Methods for Wind Farm Wake and Their Applications [Doctoral dissertation, North China Electric Power University (Beijing)].
4. Li, L., Liu, Y. Q., Yang, Y. P., et al. (2013). Short-Term Wind Speed Prediction Method Based on CFD Flow Field Pre-Calculation. Proceedings of the CSEE, 33(07), 27–32+22.
5. Wang, Y. (2014). Research on power prediction methods for wind farms in complex terrain based on CFD flow field pre-computation (Master's thesis). North China Electric Power University.
6. Annoni, J., Fleming, P., Scholbrock, A., et al. (2018). Analysis of Control-Oriented Wake Modeling Tools Using Lidar Field Results. Wind Energy Science, 3(2), 819–831.
7. Liao, H. (2020). Research on Active Power Control Optimization of Wind Farms Considering Fatigue Loads [Doctoral dissertation, University of Electronic Science and Technology of China].
8. Chen, D. (2014). Offshore Wind Turbine Foundation Structures. China Water & Power Press.

Chapter 7
Fault Diagnosis and Health Management of Offshore Wind Turbines

7.1 Introduction

The cost of operation and maintenance (O&M) of offshore wind farms accounts for more than 20% of the total cost of the project over its entire life cycle, which is two to three times the cost of onshore wind power. O&M of the offshore wind power has brought huge challenges to the economics and profitability of an offshore wind farm, and have gradually become one of the main factors affecting the development of offshore wind power.

Maintenance of offshore wind farms presents entirely different challenges than onshore wind farms. This chapter introduces the O&M challenges of an offshore wind farm, which are high fault rates, poor accessibility, and high maintenance cost. Focusing on three main challenges, the key technologies of fault diagnosis of offshore wind power devices and O&M of an offshore wind farm are introduced. In the end, the technological frontiers in the near future are discussed.

7.2 Challenges of Offshore Wind Turbine Operation and Maintenance

The installed capacity of offshore wind power has grown rapidly. However, due to maintenance difficulties and high costs, the cost per kilowatt hour of offshore wind power is much higher than that of onshore wind power, which largely impacts the profitability of offshore wind farms. The special geographical location and environmental conditions of offshore wind farms bring huge challenges to offshore wind power, which are reflected in the following three aspects.

(1) Higher failure rate of offshore wind turbines

© The Author(s), under exclusive license to Springer Nature Singapore Pte Ltd. 2025
Y. Liu et al., *Offshore Wind Farm Technology*,
https://doi.org/10.1007/978-981-96-1889-7_7

Offshore wind farms have high average wind speeds and high annual available hours. The equipment used in an offshore wind farm is more susceptible to harsh natural conditions such as salt atmosphere, typhoons, sea waves, lightning, and icing. Therefore, the components of offshore wind turbines might fail quickly and their service life be shortened. In addition, because offshore wind farms are far away from the shore, it is not convenient for regular daily inspections. Therefore, the failure rate of offshore wind turbines is significantly higher than that of onshore wind turbines. According to statistics, the average annual availability rate of offshore wind turbines is only 70–90%, which is far lower than the 95–99% availability rate of onshore wind turbines.

(2) Poor accessibility of offshore wind turbines

Most of the offshore wind farms are located in areas where the oceanic climate and continental climate are alternately affected, and the weather and waves change greatly. The use of marine transportation equipment (such as maintenance vessels and helicopters, etc.) is highly limited by the weather. When the wave height or wind speed exceeds the safety threshold, technicians cannot reach an offshore wind turbine for maintenance for safety reasons. Consequently, the maintenance time is random and much shorter than that of onshore. According to statistics, at the current technical level, the accessible time to offshore wind turbines is about 200 days per year. This will decrease with the deterioration of sea conditions.

(3) Higher cost of offshore wind turbines

The maintenance of offshore wind farms needs to rent or purchase special transport ships, hoisting ships and helicopters, etc. The cost of transportation and hoisting of parts is much higher than that of onshore wind farms. In addition, the maintenance of offshore wind turbines is limited by sea conditions. Therefore, timely and effective maintenance of wind turbines is often not possible. This results in a certain amount of power loss and indirectly increases the O&M costs of offshore wind farms.

7.3 O&M and Health Management of Offshore Wind Farms

The maintenance schedule and cost of offshore wind farms will be determined by a variety of factors, including the availability of maintenance personnel and vessels, spare parts, and weather conditions. A combination of optimized O&M strategies and advanced O&M technologies (such as condition monitoring, fault diagnosis, fault prediction, etc.) can reduce the O&M costs of an offshore wind farm.

This section summarizes existing technologies and research from three perspectives: maintenance strategy, accessibility estimation, and maintenance path optimization. The current problems with offshore wind power maintenance and management technology are summarized, and future research trends are suggested.

7.3.1 Maintenance Strategy

The maintenance cost of offshore wind farms includes not only labour costs, maintenance costs, spare parts, but also the power loss caused by the downtime of wind turbines. Therefore, it is very essential to study the maintenance strategy of offshore wind farms to ensure their economics and reliability. There are many ways to classify maintenance strategies. According to the regulations of the European Institute of Standardization, the maintenance of wind turbines can be divided into three types: break-down maintenance, planned maintenance, and condition-based maintenance, as shown in Fig. 7.1.

(1) Break-down maintenance

The break-down maintenance strategy means that before the equipment fails, no preventive maintenance would be carried out until the equipment fails. Due to the harsh offshore environment, complex geographical locations and difficulty in transportation conditions, compared with the onshore wind farm, maintenance personnel will be more difficult to approach faulty equipment. In the face of a random failure, if it cannot be repaired in time, it will lead to longer downtime and a huge loss of power generation.

For offshore wind farms, although break-down maintenance can reduce the frequency of maintenance and inspection of wind turbines to a certain extent, it will lead to longer maintenance time and downtime. Therefore, such strategy is mainly

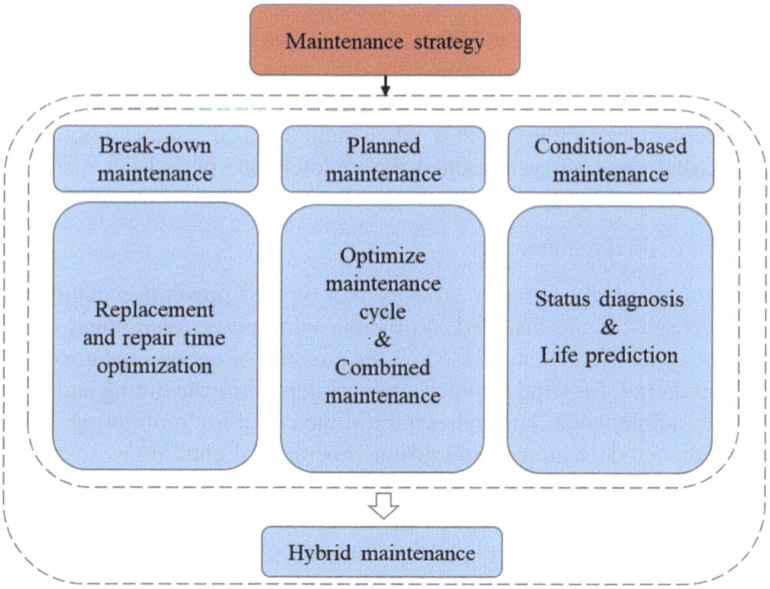

Fig. 7.1 Classification of offshore wind maintenance strategies

applicable to components with low importance and low maintenance costs, while its economics can be far lower than other types of maintenance strategies. According to statistics, the break-down maintenance costs of existing offshore wind farms account for 65–75% of the total O&M costs.

(2) Planned maintenance

The planned maintenance strategy means that on the basis of a certain understanding of the failure rules of the equipment, no matter what the health condition of the equipment is, maintenance is carried out according to the predetermined time. It is currently the most economical and feasible maintenance strategy, and it is also the most widely used maintenance strategy.

The planned maintenance is mainly divided into daily inspections and special inspections. The daily inspection of offshore wind farm equipment mainly conducts tour inspections on wind turbines, foundations above the water surface, offshore booster stations, wind measurement devices, and on-site high-voltage power distribution lines. The special inspection is an additional and special tour inspection in the case of storm surges, typhoons, abnormal hydrometeorological conditions at sea, or after abnormal operation of wind turbines and/or offshore booster stations, emergency repairs (or major repairs) of wind turbines, and new equipment put into operation.

The research related to the planned maintenance strategy of offshore wind farms includes planned maintenance strategy optimization and combined maintenance based on planned maintenance strategy. Research on the optimization of planned maintenance strategies has mainly focused on optimizing the planned maintenance cycle. If the planned maintenance cycle is not selected properly, over-maintenance or under-maintenance will occur, resulting in high maintenance costs or low reliability.

The combined maintenance strategy is an extension of the planned maintenance strategy. In order to reduce costs, some scholars have recently proposed the combined maintenance method, that is, when maintaining a certain component, other components that have not yet reached the maintenance period are maintained in advance.

(3) Condition-based maintenance

The condition-based maintenance strategy is a type of preventive maintenance in which various sensors are installed in offshore wind power equipment in order to collect current operational data. Based on the analysis of operational condition data and health prediction for wind turbines, workers determine the timing and content of wind turbine maintenance. Through wind turbine condition monitoring, condition-based maintenance determines wind turbine operational conditions, so as to detect faults in time and formulate correct and effective maintenance plans. Condition-based maintenance can make the maintenance management of wind power equipment gradually transition from planned maintenance and break-down maintenance to preventive maintenance based on condition monitoring. Figure 7.2 shows the schematic diagram of condition-based maintenance. Condition-based maintenance

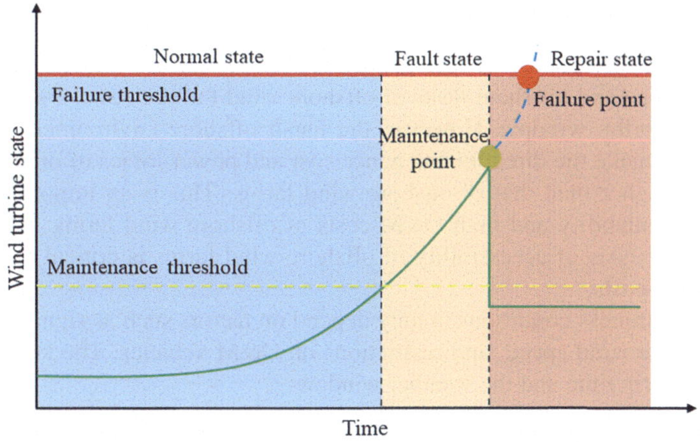

Fig. 7.2 Schematic diagram of condition-based maintenance

can maximize the reliability of wind turbines, reduce unnecessary maintenance, and reduce downtime and maintenance costs.

Condition-based maintenance requires complex and expensive condition-monitoring equipment and systems. Due to cost considerations, relatively few offshore wind farms currently in operation use condition-based maintenance. Technological advances will continue to reduce the cost of condition monitoring systems, and the research results of condition-based maintenance strategies for offshore wind farms will be crucial to their future development.

For condition-based maintenance, the selection of the maintenance threshold will have a greater impact on the maintenance cost. When an early failure of the wind turbines or components appears, predicting the remaining life of wind turbines or components provides significant guidance for the O&M of offshore wind farms. The prediction of the remaining life of wind turbines or components based on condition monitoring can arrange maintenance plans in advance and allocate maintenance resources to avoid losses caused by the failure of wind turbines or components.

(4) Hybrid Maintenance Strategy

Each maintenance strategy is suitable for specific components. A single maintenance strategy cannot meet the demand for the lowest overall O&M costs for wind turbines. At present, a large number of studies have begun to use a combination of multiple maintenance strategies to minimize total O&M costs. In the future, the research direction of hybrid maintenance strategies is to choose different maintenance strategies for different offshore wind power equipment to minimize comprehensive O&M costs of wind farms, while meeting the reliability requirements of wind turbine.

7.3.2 Accessibility Estimation

The maintenance and overhaul tasks of offshore wind farms need to be carried out within a seaworthy window. However, the harsh offshore environment and poor accessibilities make the direct maintenance cost and power losses of offshore wind farms much higher than that of onshore wind farms. This is an important reason for the low availability and high O&M costs of offshore wind farms. As a result, quantitative analysis of accessibility of offshore wind farms is critical to reducing their life cycle costs.

The seaworthiness conditions mainly depend on factors such as significant wave height, offshore wind speed, and restrictions on O&M vehicles. The key elements are the round-trip time and the weather window:

- Round-trip time refers to the sailing time of the ship from the dispatch centre or the previous task execution point to the location where the offshore wind power equipment needs to be maintained, and the execution time of the maintenance work. Generally, the larger the wind farm and the farther it is from the shore, the longer the round-trip time. The more complex the maintenance task that needs to be performed, the longer the required round-trip time.
- The weather window refers to the time period when the O&M vessels can safely arrive at the maintenance site and perform maintenance work according to the wave height and weather conditions. The higher the wave height that the O&M ship can withstand, the longer the possible weather window. Some literature shows that the significant wave height limit for O&M vessels is usually 1–3 m, and the wind speed limit is usually 10–20 m/s. Helicopters have no wave height limit, and the wind speed limit is 20–40 m/s.

Zhang Hao et al. from North China Electric Power University proposed a multi-step probabilistic prediction model for the significant wave height of the next 1–12 h and an uncertainty simulation method for the accessibility of offshore wind farms. The author proposes a probabilistic prediction model using sequence-to-sequence methods in deep learning. The O&M process of offshore wind farms is simulated by the Monte Carlo method. The time delays in the O&M process are divided into three categories, and the frequency and time distribution of the three types are calculated. Furthermore, the accessibility probability and the total time delay distribution of O&M when selecting different O&M vessels are analysed, to guide maintenance strategies.

7.3.3 Maintenance Path Optimization

In offshore wind farms, multiple wind turbines may fail at the same time. In the face of expensive leasing and using fees of offshore transportation equipment, how to complete all maintenance tasks efficiently, quickly, and economically within the limited seaworthiness window period is another difficult problem faced by offshore wind power maintenance management. Specifically, it includes reasonable planning of maintenance routes and dispatching of personnel, vessel configuration to transport technicians, equipment, and spare parts, etc. to the maintenance site through O&M vessels or helicopters.

The path planning of O&M vessels is a complex multi-constraint combinatorial optimization problem. At present, the research on maintenance route optimization of offshore wind farms can be divided into single wind farm maintenance route optimization and wind farm clusters maintenance route optimization.

O&M service providers can use multiple ports as O&M bases to provide O&M support for multiple offshore wind farms. With the development of more and more clusters of adjacent wind farms, it is more and more important to consider the maintenance path optimization and scheduling of multiple O&M centres and multiple wind farms. This method can provide a solution to reduce the O&M costs of large offshore wind farms.

7.4 Fault Diagnosis of Offshore Wind Power Equipment

The core of condition-based maintenance is the condition monitoring and fault diagnosis technology of wind power equipment and core components. Condition monitoring and fault diagnosis technology are applicable to key components with degraded failure processes such as gearboxes, main shafts, generators, and blades, as well as electrical equipment in offshore wind farms.

For different monitoring signals, different fault diagnosis techniques have been developed. The signals that can be monitored include vibration signals, acoustic signals, electrical signals, temperature signals, oil components, etc. Various sensors are installed on offshore wind power equipment, which can measure a large amount of condition data. Since offshore wind power equipment often operates under harsh wind and sea conditions, and the operational conditions of the equipment itself are relatively complex, the collected signals often show strong non-stationarity and nonlinearity and sometimes have a low signal-to-noise ratio. How to process and analyse these signals is a key problem in condition monitoring and fault diagnosis of wind power equipment.

7.4.1 Classification of Fault Diagnosis Techniques for Offshore Wind Farms

1. **Classified by monitoring equipment**

Offshore wind farms consist of numerous pieces of equipment. Different from the research object of onshore wind farm fault diagnosis technology, the research object of offshore wind farm fault diagnosis technology not only focuses on offshore wind turbines and their key equipment but also needs to focus on offshore power transmission equipment, such as submarine cables.

Fault diagnosis and health management need to focus on components with high failure rates or long downtime. Due to the different types and structures of wind turbines and their high reliability, it is difficult to collect data for various faults and their corresponding downtime in a short period. Therefore, in order to obtain a more accurate failure rate, it is necessary to cover various types of wind turbines as much as possible, a sufficient number of wind turbines for each type, and last for a long enough time. Figures 7.3 and 7.4 respectively show the failure rate and downtime statistics of each component of onshore and offshore wind turbines. The results are based on 18 public reliability datasets of wind turbines around the world, including 14 datasets of onshore wind turbines and 4 offshore wind turbines. Specifically, it includes the failure rate and downtime of more than 90,000 operating hours of 18,000 wind turbines. It can be seen from the figure that the failure rate and downtime of most components of offshore wind turbines are much higher than those of onshore wind turbines. At the same time, the downtime caused by gearbox and generator failures of offshore wind turbines is much higher than other components. The main cause of gearbox and generator failures is gearbox and generator bearing failures. Gearboxes and generators do not have the highest failure rates, however, due to the difficulty of repairing these components, the resulting downtime is the longest. Therefore, the health of bearings, generators, and gearboxes is the most critical concern.

2. **Classified by condition monitoring technology**

(1) Vibration monitoring technology

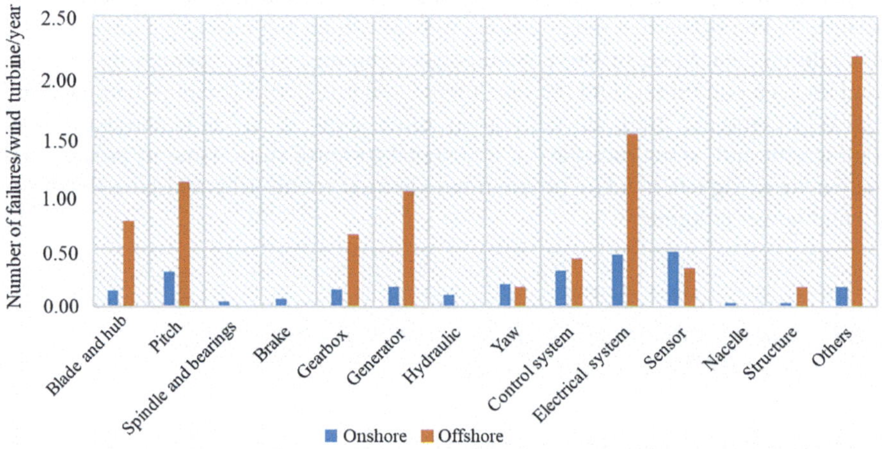

Fig. 7.3 Failure rate statistics of components of onshore and offshore wind turbines

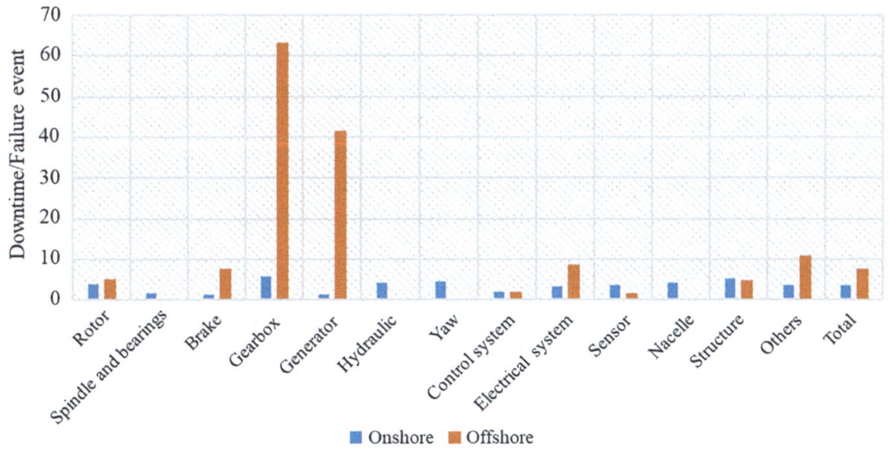

Fig. 7.4 Downtime statistics of components of onshore and offshore wind turbines

Vibration monitoring technology is the most widely used technology in the field of wind turbine condition monitoring. It can cover the condition monitoring of gearboxes, bearings, generators, nacelles, blades, towers, foundations, and other components of wind turbines. Figure 7.5 is a schematic diagram of the layout of vibration monitoring locations.

Vibration signal analysis is to process and analyse the extracted vibration signals, and then dig out the fault information of wind turbines hidden in the vibration signals. Commonly used methods are time-domain analysis, frequency-domain analysis, and time–frequency analysis.

1) *Time domain analysis*

The time-domain analysis method refers to the analysis of characteristics in the time-domain waveform such as amplitude, period, shape, square root, and kurtosis.

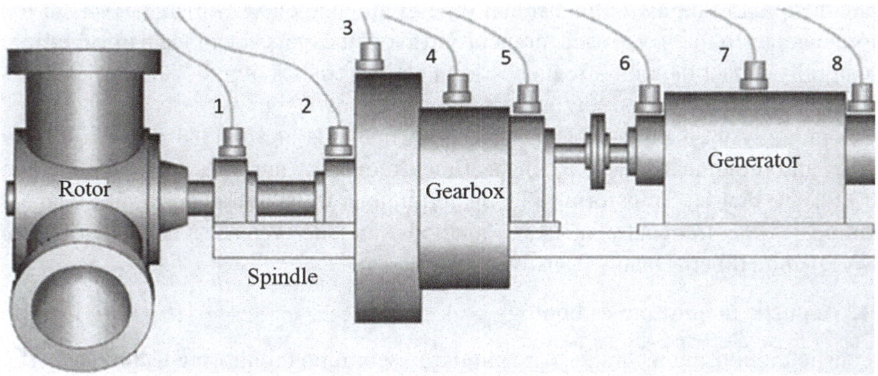

Fig. 7.5 Schematic diagram of the layout of vibration monitoring locations

Although this method has the advantages of simplicity and convenience, it also has obvious limitations. First, the time-domain waveform of the vibration signal has no obvious rules when the wind turbine is running under some operational conditions, and effective information cannot be obtained directly from the time-domain waveform. Second, when the signal is weak, it may be interfered by other noise signals, resulting in low accuracy of fault diagnosis. Therefore, the time-domain analysis method is generally used as a preliminary diagnostic for vibration signals.

2) *Frequency domain analysis*

Frequency domain analysis is the most commonly used method in vibration signal processing. The method is based on Fast Fourier Transform (FFT), including envelope analysis, backstep spectrum analysis, and spectral kurtosis method. The Fourier transform method decomposes complex signals into a harmonic component of a limited frequency, and realizes the transition from the time domain to the frequency domain.

When a wind turbine fails in operation, new frequency information will appear in its frequency spectrum diagram or the amplitude of the original signal frequency will change. This indicates a new failure or an exacerbation of an existing failure. Compared with the time-domain analysis method, the frequency-domain analysis method is more intuitive and is usually used for vibration signal analysis of bearings, gears. Due to changes in speed, load, and working environment, the vibration characteristics of wind turbines are nonlinear and non-stationary, so the frequency domain analysis method cannot be accurately processed. For such non-stationary signals, more advanced signal processing methods are required.

3) *Time–frequency analysis*

Time–frequency analysis is the abbreviation of time–frequency combined domain analysis. This method can provide joint distribution information of the time domain and frequency domain. It can clearly describe the relationship of the signal frequency over time and the vibration density and intensity at different times and different frequencies. The time–frequency analysis method maps the one-dimensional time-domain signal to the two-dimensional time–frequency plane, which can obtain the instantaneous frequency at each moment and avoid the omission of local information. It can fully reflect the time–frequency joint characteristics of non-stationary signals and is an effective method for analysing non-stationary signals.

Commonly used time–frequency analysis methods are divided into two types: linear and nonlinear. Typical linear time–frequency analysis methods include Short-Time Fourier Transform (STFT), continuous wavelet transform, etc. Typical nonlinear time–frequency analysis methods include Wigner-Ville Distribution (WVD) and Hilbert-Huang Transform (HHT), etc.

(2) Acoustic monitoring technology

Acoustic monitoring techniques are similar to vibration monitoring techniques. The difference between the two is that the vibration sensor is rigidly installed on the part to be monitored, and the acoustic sensor is pasted on the part to be monitored by a flexible adhesion with low attenuation. The most effective bearing condition monitoring technology based on acoustic measurements is the acoustic emission technology. The most important application of acoustic emission technology is the detection of cracks.

(3) Electrical signal monitoring technology

Electrical signal-based condition monitoring typically detects abnormal electric components based on signals such as voltage and electric current. Since electrical resistance varies with the stiffness of the components, it is possible to detect changes in stiffness that may be caused by cracks, delamination, or fatigue.

(4) Oil monitoring technology

The important information on wind turbine will be reflected in the changes of various indicators in the lubricating oil used. It is possible to determine the operational condition and to diagnose the type, location and reasons of wearing failure through monitoring and tracking the states of lubricating oil of main gearbox, pitch system, yaw system, brake control device such as lubricating oil physics and chemical status, wear metal particles and pollution impurities particles, etc.

Many wind farm O&M personnel send oil to designated commercial laboratories for inspection to monitor equipment operational and fault conditions. However, depending solely on laboratory test reports will significantly reduce the effectiveness of oil monitoring. The main reasons for this are in three aspects. First, the timeliness is poor; commercial laboratories usually issue reports within 1–4 weeks, and it will take even longer time to deliver samples and transmit reports. Second, the wind farm is located in a remote area, where the cost of delivery is high. The transportation costs will be significantly higher if oil samples are frequently sent to the laboratory for analysis. Moreover, the delivery control of liquid items is becoming more and more strict, and it is more likely to encounter situations where oil samples cannot be delivered. Third, relevant inspections require professional equipment and personnel, and the level of inspection directly affects the accuracy of fault diagnosis.

In order to solve the problem of timeliness, real-time oil monitoring can be used, which can effectively manage maintenance and lubrication, and avoid major accidents. It has been recognized by more and more wind farm O&M personnel and wind farm owners.

2. **Classified by the signal source**

(1) Based on the CMS system

Condition Monitoring System (CMS) is a monitoring system for equipment condition parameters (such as temperature and vibration). The purpose is to determine whether there is an abnormality in the equipment according to the parameter change, generally the large changes.

Currently, CMS has been successfully applied to the wind power industry. CMS for many wind turbine components have been widely used, including gearboxes, lubricating oil, blades, towers, and foundations. For example, the health conditions of wind turbines can be reflected by monitoring the vibration trend of the drive chain. The operational conditions of the gearbox can be reflected by monitoring relevant parameters such as the moisture particles in the lubricating oil. The health conditions of the blade can be determined by monitoring the vibration displacement or sound frequency and decibel value when the blade is running. Tower shaking and uneven settlement can be monitored through the online system.

Although CMS has the advantages of high precision and strong pertinence, it still faces some challenges when it is applied in the field of offshore wind power. Firstly, for offshore wind power equipment in a marine environment, the corrosion rate is usually faster than that on land. These effects are further exacerbated by higher wind speeds and poor accessibility at sea, leading to longer downtime. Current commercial wind turbine CMS has not considered the detection, diagnosis, and prognosis of such conditions. Secondly, CMS needs to install additional sensors for each monitoring object. The number of wind turbines in offshore wind farms is large, the reliability requirements are high, and the investment cost is very high. Finally, CMS consists of electronic components and sensors that are particularly prone to failure in harsh offshore environments. Regular maintenance and recalibration of the CMS are very significant but also bring high maintenance costs. At the same time, operators are also faced with the challenge of processing, transmitting, storing, and interpreting large amounts of monitoring data.

(2) Based on the SCADA system

SCADA system is a monitoring system used to ensure the safe and stable operation of wind turbines. Since the SCADA system will record the key data and alarming condition of the wind turbine operation, the SCADA data of the wind turbine can be used as critical data for fault diagnosis. The biggest advantage of this system is that it does not require any additional sensors or hardware.

However, compared with the specially developed condition monitoring system, the monitoring and diagnosis system based on the SCADA system has considerable limitations and cannot replace the professional CMS:

- The SCADA system does not collect complete condition monitoring signals from wind turbines;
- The data in the SCADA system is recorded at a very low sampling rate (usually 10 min or 15 min), which cannot meet the needs of some condition monitoring and fault diagnosis;
- Data analysis tools based on SCADA systems have not been widely validated, and because wind turbines operate under time-varying conditions, SCADA systems often give false alarms;
- Health condition indicators based on SCADA data, such as bearing temperature rise, are generally considered to be only late fault diagnosis results, which means that when some abnormalities are found, it is usually time to shut down the turbine and is unable to leave necessary failure prognosis time.

7.4.2 *Fault Diagnosis of Key Components of Offshore Wind Turbines*

This section will specifically introduce the vibration characteristics of key components of wind turbines such as gearboxes, bearings, generators, and blades when they fail, as well as other effective fault diagnosis methods applicable to these key components.

1. **Gearbox fault diagnosis**

Vibration analysis is one of the most widely used methods for diagnosing gearbox problems, and the key is identifying the vibration characteristics under typical fault conditions. The vibration of the gearbox mainly comes from the meshing of the gears. The principal frequencies are the rotation frequency of each shaft, the meshing frequency of the gears, the natural frequency of the gears, and various sidebands generated by the modulation of these frequencies. No matter whether the gear is running normally or not, it will generate its own natural frequency vibration, and the vibration pattern may be different. The natural frequency vibration of the gear is mostly high-frequency vibration at about 1 ~ 10 kHz. When the high-frequency vibration is passed to the gearbox, the vibration impact has been attenuated, and in most cases, only the meshing frequency of the gears can be monitored. The vibration characteristics will be introduced from the three aspects of normal meshing, gear wear, and local failure of the gear.

(1) Vibration characteristics under normal meshing state

Normally meshed gears vibrate more smoothly. The time-domain waveform is a periodic attenuation waveform, and the low-frequency signal is similar to a sine wave. The normal gear vibration usually contains both the meshing frequency and its harmonic frequency, with the meshing frequency being the predominant frequency. There is a decrease in the higher harmonic frequency. As a result, there will be a rotation frequency of the gear shaft and its higher harmonic frequency at low frequencies.

(2) Vibration characteristics under gear wearing state

Gear wear is usually caused by abrasive wear. When the gear tooth surface is worn, the position of the meshing frequency and its harmonic component on the spectrogram remains unchanged. However, the amplitude changes and the amplitude of the higher harmonic frequency increases relatively. Therefore, the main characteristic of gear wear is the change in meshing frequency and the amplitude of its harmonic component. In addition, with the development of wear, the stiffness of the gear shows the characteristics of non-linearity; the vibration waveform will appear with higher harmonics of the meshing frequency with 2 or 3 times of meshing frequency or the fractional harmonics of the 1/2 or 1/3 meshing frequency; or the phenomenon of amplitude jump with non-linear vibration characteristics appears.

(3) Vibration characteristics of Gears under partial fault state

Local faults in gear include large cracks, local wear of the tooth surface, local errors in the tooth shape, and broken teeth. When the gear operates with local abnormalities, the mesh of the abnormal parts will have a great impact. The number of impacts depends on the number of local failures. In the time domain, the gear under this condition usually exhibits a large amplitude and regular shock vibration. The frequency of this shock vibration is usually equal to the rotational frequency of the shaft on which the faulty gear is located. In the frequency domain, the specific manifestations of local abnormal gears are as follows. Firstly, sidebands with intervals of shaft rotation frequency appear near the meshing frequency and its higher harmonic frequency. Secondly, signals above the 10th order may appear at the rotation frequency and its higher harmonic frequency. Thirdly, due to the large impact energy, the natural frequency of the gear will be excited, therefore, there will be a modulation sideband, whose carrier frequency is the natural frequency of the gears and its modulation frequency is the rotation frequency of the fault gear and its high harmonic frequency.

2. **Bearing fault diagnosis**

As a crucial part of wind turbines, bearings are used in the main shaft, gearbox, and generator in the drive train. The bearings in the drive train are usually rolling bearings. The vibration of such bearings is mainly caused by the contact with the inner and outer ring. The vibration characteristics of the bearings will be introduced from three aspects, which are normal operation, bearing wear, and local faults of the bearings.

(1) Bearing vibration characteristics under normal operating conditions

Complex vibrations are also generated during normal bearing rotation. These vibrations mainly include the natural vibration of the bearing and the vibration excited by an external force, as well as the vibration caused by the elastic characteristics of the bearing. The stiffness of the rolling body is very large. Under the action of load, generally, only a small elastic deformation will occur. When the load is too large, nonlinear elastic deformation and nonlinear vibration will occur.

(2) Vibration characteristics under bearing wear condition

As the bearing rotates and the wear increases, the acceleration peak value and root mean square (RMS) value of the vibration gradually increase, and the vibration signal will show strong randomness. If fatigue spalling does not occur, the final vibration amplitude will be many times larger than the initial one.

(3) Vibration characteristics of the bearing under local fault conditions

If there are local defects (such as spalling, pitting, cracks, indentations, etc.) on the rolling body of the bearing or the inner and outer rings, impacts will occur when the contact points pass through the defects. The vibration frequency caused by periodic impact depends on the location of the fault, so it is called the fault pass frequency, and the calculation equation is as follows.

The Ball Spin Frequency (BSF) is shown in Eq. 7.1.

$$f_B = \frac{D}{d}\left(1 - \left(\frac{d}{D}\cos\varphi\right)^2\right)f_n \tag{7.1}$$

The Ball Pass Frequency Inner race (BPFI) is shown in Eq. 7.2.

$$f_I = \frac{1}{2}Z\left(1 + \frac{d}{D}\cos\varphi\right)f_n \tag{7.2}$$

The Ball Pass Frequency Outer race (BPFO) is shown in Eq. 7.3.

$$f_O = \frac{1}{2}Z\left(1 - \frac{d}{D}\cos\varphi\right)f_n \tag{7.3}$$

The Fundamental Train Frequency (FTF) is shown in Eq. 7.4.

$$f_F = \frac{1}{2}\left(1 - \frac{d}{D}\cos\varphi\right)f_n \tag{7.4}$$

where d is the diameter of the rolling body; D is the pitch diameter of the bearing; Z is the number of rolling bodies; and φ is the contact angle.

3. **Generator fault diagnosis**

This section describes the mechanical (mainly vibration) and electrical (mainly electric current) characteristics of a failed generator.

(1) Generator vibration characteristics

The main electromagnetic vibration forms of the generator include electromagnetic vibration generated by radial electromagnetic force and fundamental magnetic field. In addition to electromagnetic vibration, there will be other forms of vibration, including mechanical vibration of the rotor and bearing, and vibration generated by devices such as commutators and electrical brushes.

Due to the mass imbalance of the generator rotor, centrifugal force will be generated during operation. This will cause periodic structural vibration, whose frequency is mainly the rotation frequency of the rotor. In addition, if the rotor has other installation problems, such as shafting misalignment, rotor bending, etc., it will also cause vibration. There will be additional abnormal vibrations if there is a fault with the rotor, such as cracks, falling parts, looseness of the iron cores or coils, broken bars, etc. There are differences in vibration frequencies generated by different faults, according to this fact, the rotor faults can be diagnosed.

The rotor of the generator is supported by bearings at both ends. The supporting bearings of the generator rotor of wind turbines are usually rolling bearings. The vibration generated by rolling bearings during operation mainly comes from the excitation generated by the contact between the rolling bodies and the inner/outer rings. As a result of errors during manufacturing and installation, the inner and outer rings of the bearing have ripples or the size of the rolling elements is uneven, which will result in vibration of the bearing. Under normal working conditions, because the number of rolling bodies carried by different parts is different, the load-bearing stiffness of the bearing changes, causing the shaft centre to fluctuate and vibrate. For heavy-duty bearings, the contact between the rolling bodies and the inner/outer rings may cause deformation and vibration. Additionally, if local faults occur on the inner and outer rings of rolling bearings or rolling bodies (such as spalling, pitting, cracks, indentations, etc.), impact excitation will be generated once the fault point passes through the stressful region, causing additional periodic impact vibrations and natural vibrations of the structure. The impact vibration generated under the fault condition is periodic, and its frequency depends on the location of the fault, which is called the fault pass frequency.

(2) Generator rotor fault current characteristics

Besides vibration characteristic analysis, electrical characteristic analysis can also be used to diagnose generator faults. Due to the convenient installation of the current sensor, the stator current signal is easy to collect. Therefore, it is a very effective method to use stator current characteristics to analyse the fault of the generator.

The analysis of the stator current signal mainly adopts the frequency domain analysis method. Ideally, the stator current has only a fundamental frequency component. When the rotor circuit is faulty, there will be a sideband on the frequency spectrogram of the stator current, and its amplitude is one of the effective fault features.

Broken rotor bar fault is a frequent fault in wind turbine Doubly-fed induction generators (DFIG). When a broken rotor bar fault occurs, the rotor current will generate two rotating magnetic fields with equal magnitude and opposite directions. In the frequency of the stator current signal, the broken rotor bar fault will generate a sideband with frequency $(1-2s)f_0$. This sideband will cause periodic changes in the amplitude of the stator current signal, and generate a torque disturbance ΔT and a rotational speed disturbance $\Delta\omega_r$ with a frequency of $2s$ on the rotor. The speed disturbance can generate three induced electromotive forces with frequencies f_0 and $(1\pm 2s)f_0$ in the symmetrical three-phase stator windings. The electromotive force component of frequency $(1-2s)f_0$ will generate current I_{2s}', which interacts with I_{2s}. The electromotive force component with frequency $(1+2s)f_0$ will generate current I_{2s}'', and will generate a magnetic field with frequency $3sf_0$. Because the rotor resistance is unbalanced, two rotating magnetic fields with a frequency of $\pm 3sf_0$ will be formed. The newly generated $-3sf_0$ rotating magnetic field will excite a three-phase current component with a frequency $(1-4s)f_0$ in the stator winding current, thereby generating new rotor ripple torque and ripple speed. By analogy, the frequency component of the stator current increases when the rotor winding fails. The added frequency components are as follows:

$$\begin{aligned} &f_0 \rightarrow (1-2s)f_0 \rightarrow (1+2s)f_0 \rightarrow (1-4s) \\ &f_0 \rightarrow (1+4s)f_0 \rightarrow (1-6s)f_0 \ldots \end{aligned}$$

where $(1-2s)f_0$ is the first sideband, which has a larger amplitude, and the other sidebands have smaller amplitudes. The wind turbine generator fault diagnosis method based on electrical characteristics mainly monitors the first sideband amplitude of the generator stator current signal.

4. **Blade fault diagnosis**

The blades of wind turbines are usually made of glass fibre as the enhancement and epoxy resin as the substrates. It is made through multi-layer paving. During the manufacturing and service process, various types of damage will inevitably occur, mainly including fatigue, decreased stiffness, cracks, increased surface roughness, lightning, icing, and deformation, etc.

(1) Diagnosis method based on modal data

The damage of wind turbine blades may cause the reduction of structural stiffness and the change of modal parameters such as eigenfrequency and frequency response function. The diagnosis method based on modal data utilizes this mechanism for damage identification and diagnosis, which is simple, effective, and low-cost. Types of modal data include eigenfrequency, modal shape, curvature, strain energy, flexibility, damping, sensitivity, etc. Based on different types of modal data, diagnosis methods can be divided.

(2) Diagnostic method based on static parameters

Structural model data is a reflection of structural integrity. The static parameters such as strain have the ability to describe the local conditions of the structure and are more sensitive to the local damage of the structure. The static parameters can be used to identify structural damage based on the differences in strains and displacements when the structure is damaged or not. Among these methods, damage identification methods based on strain parameters are widely used. In the damage identification method based on strain parameters, accurate measurement of strain parameters of each measurement point of the structure is an important basis for accurate damage identification. Strain parameter measurement sensors mainly include strain gauges and Fibre Bragg Gratings (FBG). Among them, strain gauges are low in cost, small in mass, and can be firmly bonded, which is a traditional way of measuring strain parameters. However, when there are many measurement points, a large number of connecting lines need to be attached to the structure under test. This would have a non-negligible impact on the actual state of the measured structure, which makes high-precision measurement more difficult. FBG is an emerging method for measuring strain parameters. The basic principle is that the use of stress changes can cause changes in the grating spacing and refractive index in the FBG, which has the advantages of high measurement accuracy, enhanced sensitivity, small mass, and is easy to achieve multi-point measurement. However, the drawback is that a high-cost grating demodulator is needed.

(3) Diagnosis method based on vibration signal

The diagnosis method based on the vibration signal is to use a piezoelectric ceramic sensor to capture the vibration signal. There are four main methods for blade fault diagnosis. First, the transfer function and dynamic deformation analysis methods need to use Scanning Laser Doppler Vibrometers (SLDV) and the measurement data of the blade in a healthy condition as a reference. Therefore, it is difficult to apply in practice although the diagnosis results are more accurate. Secondly, the response comparison and wave propagation analysis method only require piezoelectric ceramic sensors and vibration exciters. Thirdly, the wave propagation method is only sensitive to faults located between the sensor and the vibration exciter, which has limitations. Fourthly, the response comparison analysis method is simple and has low requirements for historical data.

(4) Diagnosis method based on electromechanical impedance

Damage can cause changes in the impedance of the piezoelectric (PZT) patch in the structure, and damage recognition can be achieved according to the principle. The method based on electromechanical impedance is an active structural health monitoring method and integrates signal excitation and signal collection. This method has the advantages of high local sensitivity, simple and convenient system integration, no need for model analysis, and is especially suitable for online monitoring of flat-plate structures. This method does not require a model analysis of the structure, but an appropriate mapping model. This mapping model quantifies the relationship between the impedance signal and the physical parameter to be measured.

In the electromechanical impedance diagnosis method, the obtained impedance signal is used to calculate the mechanical impedance of the structure in question. The damage can be identified and located through the change in impedance in the structure. The method uses piezoelectric patches as excitation/sensors to obtain impedance signals. However, piezoelectric materials are easily affected by ambient temperature. Therefore, to improve the robustness and effectiveness of such methods, a certain measure of temperature resistance is needed. For example, artificial intelligence methods, such as neural networks, can compensate for temperature and adverse responses. At the same time, since this method is currently unable to detect damage at a position far from the sensor. Therefore, a large range of damage positioning cannot be achieved in the structure and the damage recognition accuracy is limited. Also, it is difficult to detect small-scale damage with high precision.

(5) Diagnostic method based on guided wave

The guided wave is a type of elastic wave including surface waves, Lamb waves, and Interfacial waves. The guided waves have aroused the interest of researchers in the field of structural health monitoring because of their sensitivity to different types of damage, long propagation distance, low energy attenuation, and wide detection range. Using guided waves for damage monitoring is an active structural health monitoring method. Compared with passive methods, it is especially suitable for online monitoring of structures. This type of method usually uses PZT or Macro Fibre Composite (MFC) films as an actuator and sensor. When there is damage in the structure to be tested, the guided wave will interact with the damage to produce a wave reflection or scattering effect, which will change the propagation characteristics and energy distribution of the guided wave. The health condition can be assessed based on these changes detected by pre-positioned sensors. In addition, the excitation frequency of this type of method is high, reaching the kilohertz or megahertz level, so that the wavelength is relatively short and small-sized damage can be identified.

Although the method based on guided waves has the advantages of being able to detect flaws on the structural surface and inside and being sensitive to small damages, it also has some disadvantages. Due to the extremely complex propagation of guided waves, especially in composite materials such as blades, the propagation process is difficult to solve accurately and quickly. Due to the dispersion and multi-mode effects of guided waves, it is necessary to combine advanced signal processing methods to extract damage-related features. Because it is still difficult to realize the damage location without a reference, it is necessary to obtain the guided wave signal when the structure is in healthy condition as the reference signal. Therefore, this direction still needs further research.

(6) Other non-destructive testing methods

Some researchers and engineers have been using Modern Non-Destructive Testing (NDT) method to identify the health condition of the blades. For example, the acoustic emission detection method uses the strain wave released when the object is fractured and deformed due to stress concentration inside the object to identify the abnormality of the inspected part. During high-frequency alternating impact and corrosion, the wind turbine blade is easy to suffer abnormalities such as cracks and deformations, which can be detected by acoustic emission. Another infrared imaging detection method is to utilize infrared imaging of objects at different temperatures to identify abnormal conditions. Since the health condition of the object surface (cracks, spalling, etc.) will affect the energy distribution of thermal radiation, infrared imaging detection methods can be used for the diagnosis and identification of cracks on the blade surface. The above-mentioned advanced methods are still mainly in the experimental stage and will take a certain amount of time to apply in practice.

7.4.3 Fault Diagnosis of Offshore Wind Power Transmission Systems

The offshore wind power transmission system required by offshore wind farms includes electrical collector systems, offshore substations, and subsea cables, etc. This section will introduce cable fault diagnosis technology and substation fault diagnosis technology.

1. **Cable fault diagnosis**

(1) Types of cables for offshore wind farms

① *Electrical collection cable*

The offshore electrical collector system connects the wind turbines inside the wind farm through the subsea cables to collect the power generated by each wind turbine before being transmitted to the onshore substation. The operating voltage is medium voltage level, generally 33 kV.

② *Infield cables*

The infield cables of the offshore wind farm are used for power transmission between the offshore power collection system and the substation, as well as between various offshore power collection systems. The increase in distance from the coast to the farm leads to higher transmission losses. To reduce power transmission loss, some wind farms use an HVDC transmission system to transmit power to land.

③ *Transmission cable*

The transmission cable is used to connect the offshore substation with the onshore substation, to transmit power with minimum loss. According to the distance from the shore, the total power generation of the wind farm, and the maintenance cost, a single-core high-voltage DC cable or a three-core (single-core) high-voltage AC cable can be used. When the transmission distance is long, high-voltage DC cables will have a great advantage in reducing power losses. When using high-voltage DC cables for transmission, substations are required both at sea (AC/DC) and on land (DC/AC), which is expensive for their construction.

(2) Types of cable faults

Faults of the subsea cable are mainly divided into electrical faults and mechanical faults. The reasons for electrical faults include aging of insulating materials, overloading, etc. The main causes of mechanical faults are human factors such as ship anchoring and natural factors such as changes in the earth's crust.

① *Electrical faults*

The electrical faults of subsea cables are divided into short-circuit faults and electric leakage faults. Leakage faults are the most common type of fault. A short circuit fault is a short circuit between supply conductors or between a supply conductor and the seabed, causing the conductor to generate heat in an instant and burn the insulation layer, and the system stops working. Leakage fault refers to the fact that the current in the conductor forms a circuit with the seawater when the insulator is damaged but the power supply conductor and optical fibre are not damaged. At this time, the potential at both ends of the subsea cable changes, but the system continues to operate.

② *Mechanical faults*

The main cause of mechanical failure is ship anchoring. There are mainly two situations. One is that when dropping the anchor, the anchor hits the cable directly, causing the cable to deform or even break. The other is that when the anchor is lifted, the anchor hooks the cable, causing the cable to bend or even break. According to statistics, ship anchoring is the primary mechanical cause of subsea cable failures. The planning, design, installation, and maintenance guide of the IEEE subsea power cable system lists anchoring damage to the first man-made submarine cable disaster.

(3) Main fault diagnosis techniques

The three-phase of the fibre composite subsea cable is paved separately, with two units in each phase. Each unit contains multiple fibre optic cables. During power transmission, some of the fibres are in standby mode. The measurement of the operational condition of the subsea cable is achieved by Brillouin optical time-domain reflection technology, which uses the sensitivity of the spontaneous Brillouin scattering effect of the fibres to temperature and strain. Depending on the detection method, Brillouin sensing systems can be divided into direct detection and outlier detection. Outlier detection is more suitable for cable fault diagnosis due to its good spectral filtering properties.

When there is an electrical problem with a subsea cable, the temperature of the sensing fibre near the problem area goes up a lot. Temperature signals can be used to monitor whether electrical failures occur in subsea cables. When a mechanical fault occurs in the cable, the shape of the cable changes, and the strain at both ends of the cable changes. Therefore, the strain signal can be used to monitor the mechanical failure of the subsea cable.

Cables operate in a complex environment and contain a large number of jamming signals. Therefore, methods such as wavelet analysis, which deals with non-smooth signals, are needed to extract fault features for fault diagnosis and fault positioning based on neural networks.

2. **Transformer substation fault diagnosis**

The goal of substation fault diagnosis is to find the sources of faults that cause protection actions and circuit breakers to trip. For example, busbar faults, transformer faults, feeder faults, etc. The fault diagnosis techniques described earlier basically follow the process of collecting monitoring signals, extracting fault characteristics, and building a diagnostic model. Substation fault diagnosis determines the fault location directly based on information from the protective switch and no longer requires signal acquisition. The more commonly used diagnostic methods are neural networks, expert systems, fuzzy theory, rough set theory, etc. Next, a substation fault diagnosis technique based on rough set theory and neural network models is introduced, as shown in Fig. 7.6.

Because each substation is built differently, it is important to first find the fault area and then the bad part. A neural network is trained for each of the fault area decision

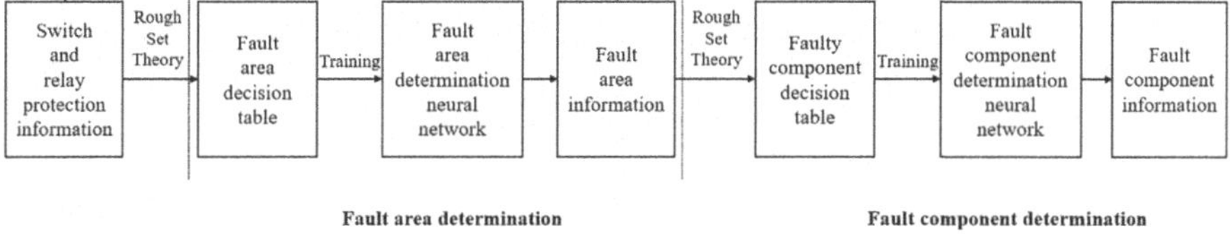

Fig. 7.6 Transformer substation fault diagnosis technology based on rough set theory and a neural network model

tables and the fault component decision table and the trained neural network can be used for fault diagnosis. The process is as follows:

- Using rough set theory to process the acquired switch protection information, remove redundant attributes, and approximate the information to form a fault area decision table;
- In the first stage of the neural network, the conditional attributes in the fault area decision table are used as inputs. The decision attributes are used as outputs to train the model. A neural network model for determining fault areas is obtained, and the corresponding fault area information is obtained;
- Fault area information and switch protection information are used as a new set of information. The same processing is done using rough set theory to form a decision table of faulty components;

The second level of the neural network gets its information from the conditional attributes of the fault component decision table. The decision attributes are used as outputs to train the model. A neural network model for determining the faulty components is built. These two layers of neural networks can be used for fault diagnosis in substations.

7.5 Offshore Wind Operations and Maintenance Technology Frontiers

7.5.1 Intelligent Diagnostic Technology

Fault diagnosis is a technology that originated in the 1960s. Its outstanding feature is the close combination of theoretical research and engineering practical application. After half a century of development, this technology has gradually matured. A relatively complete theoretical system has been formed in the aspects of signal acquisition and sensing technology, fault mechanism and symptom connection, signal processing and diagnostic methods, decision-making and diagnostic systems, and a large number of achievements have emerged. And a large number of fruitful engineering applications have been achieved in the field of fault diagnosis of offshore wind turbines. With the wide application of artificial intelligence technology, especially deep learning technology, intelligent diagnosis technology combined with artificial intelligence technology and emerging sensing technology has become an important direction for the future development of fault diagnosis technology for offshore wind turbines. It is also a direction that requires in-depth research.

Intelligent diagnosis technology is a reasoning process that simulates human thinking. Through efficiently acquiring, delivering, and processing the diagnostic information, it is capable of simulating human experts and making intelligent judgments and decisions on the operating status and faults of the monitoring objects with flexible diagnosis strategies. Intelligent fault diagnosis has a learning function and the ability to automatically obtain diagnostic information to diagnose faults in real-time. Current intelligent diagnostic technologies can generally be divided into three categories. One is the diagnosis technology based on the human experts, while the other is a based on deep neural networks and fuzzy theory.

At present, with the rapid development of artificial intelligence and deep learning, there are many new ways to make intelligent diagnoses. For example, according to expert experience and fault mechanisms, manually constructed fault features, the use of support vector machines, decision trees, and other learning methods that do not require many samples. In addition, there are also transfer diagnosis techniques using meta-learning or few-shot learning to solve the problem of lacking training samples. The core of transfer diagnosis is transfer learning, which is a new learning paradigm in machine learning. It aims to solve the machine learning task with very few or even no labelled samples in the target domain. It leverages auxiliary data and learning models to solve difficult problems with small datasets. Transfer diagnosis essentially uses the similarities between two fields to apply knowledge and experience from one field to another. One of the core problems is finding a field that is similar to the target field. When applying the transfer diagnosis method to solve the fault diagnosis problem of offshore wind turbines, it is very important to construct auxiliary data for fault samples similar to those of the questioned offshore wind turbines.

Offshore wind turbines have a trend toward large-scale, complex, automated, and intelligent development. There is an urgent need for intelligent diagnostic technology, an expert consultation platform, and remote diagnostic technology that integrate intelligent sensor networks, intelligent diagnostic algorithms, and intelligent decision-making. Different types of intelligent diagnosis methods have their advantages and disadvantages when performing fault diagnosis on a specific and relatively simple object. For the expert system diagnostic technology, it has difficulties in knowledge acquisition, lack of effective diagnostic knowledge expression, and low reasoning efficiency. For diagnostic techniques based on deep neural networks, it requires a large number of training samples, therefore bringing difficulties to obtain training samples due to factors such as resource constraints, the interests of all parties, policies, regulations, and the privacy of manufacturers of generators and gearbox. For the fuzzy fault diagnosis technology, it often needs to manually determine membership functions and fuzzy relation matrices based on prior knowledge, however, there are many difficulties in obtaining the membership function and fuzzy relationship matrix that are consistent with the real situation of the equipment.

At present, the ability to develop intelligent diagnostic methods is still relatively weak. Although there are many diagnostic methods based on deep learning, most intelligent methods need to meet certain assumptions and set certain parameters. Therefore, many fault diagnosis algorithms are verified by simulation in the research. Intelligent diagnostic methods often leave people with a black box impression, and their generalizability has not been well verified. It is difficult to meet the requirements of intelligence or to be widely employed with only several algorithms. The model performance will be greatly improved by combining several intelligent technologies with complementary performances and their advantages are complementary. Therefore, it is necessary to promote the use of existing artificial intelligence diagnostic methods and establish artificial intelligence methods based on fundamental research such as failure mechanisms. As a result, the practical technology of intelligent diagnosis and warning with rich knowledge, correct reasoning, accurate judgment, reasonable forecast, and reliable conclusion can be formed.

7.5.2 Digital Twin Technology

Digital Twin provides a real-time virtualized mapping of physical entities. The real-time data, e.g., temperature, vibration, load, etc. will be transferred into the digital twin model. Meanwhile, data representing the real environmental conditions will also be entered into the model so that the environment model of the digital twin is consistent with the changes of the actual working conditions. Through digital twin technology, fault predictions can be achieved before the device failure, so that components that are worn or in the early failure state can be replaced within the scheduled shutdown time to avoid accident stops.

In recent years, digital twins have become increasingly widespread. At the same time, thanks to the development of new generation of information technology such as Internet of Things, big data, cloud computing, and artificial intelligence, it is possible to apply digital twin technology to achieve failure diagnosis of complex wind power equipment such as gearboxes and generators.

The concept of twin originated from the Apollo plan of the National Aeronautics and Space Agency (NASA). At that time, the twin was two real physical aerospaces. Around 2003, the idea of digital twins appeared for the first time in the Full Life Cycle Management Courses of Professor GRIEVES at the University of Michigan. At that time, the term "Digital Twin" was not officially proposed. Until 2010, this term was officially proposed in NASA's technical report. In 2011, the U.S. Air Force explored the application of digital twins in the health management of the aircraft, and explored the technical challenges of implementing digital twins in detail. In 2012, NASA and the U.S. Air Force jointly published a paper on digital twins, pointing out that digital twins are one of the key technologies that drive future aircraft development. In the next few years, more and more studies have applied digital twins to many fields outside the aerospace field.

According to the definition of standardization organizations, digital twin refers to the digital expression of specific physical entities or physical processes with data connection. This data connection can ensure the convergence of the same rate between physical and virtual states, and provide integrated views of the entire life cycle of physical entities or processes, which helps optimize the overall performance. Digital twin technology has the characteristics of interoperability, scalability, real-time, fidelity, and closed-loop. Compared with simulation, in addition to using numerical calculation and solution technology, digital twins also need to rely on means of simulation, measurement, and data analysis, to perceive, diagnose and predict the state of physical entities. Compared with the Cyber Physics System (CPS), digital twins focus on the construction technology of models, and CPS focuses on system implementation.

The main technologies of digital twins include information modelling, information synchronization, information enhancement, information analysis, intelligent decision-making, information access interface, information security, etc. These technologies have achieved a lot of achievements and are still evolving rapidly. In various aspects, the rapid development of digital twin technology and related systems has been promoted, such as simulation, new data sources, interoperability, visualization, instruments, platforms, etc. With the common development of emerging technologies such as new generation of information technology, advanced manufacturing technology, and new material technology, the above elements will continue to be optimized. The development of digital twin technology will be explored and improved.

The digital twin is a technology that combines many different fields and disciplines to map real wind farms to digital wind farms. It is a master of intelligent wind power technology. Digital twin technology is still in its infancy. At this stage, digital twin technology should be used to empower the health management of high-value wind power equipment to achieve real-time online measurement services in the operating status of wind power equipment, as well as the interconnection services of physical equipment, control systems and information systems.

With the continuous progress and development of digital twin technology, it can build digital twin wind farm models that cover multi-disciplinary, multi-physic, multi-scale, multi-component, and multi-model to enhance the design, operation, maintenance, and management of wind farms. Also, the model can greatly expand the interaction process between intelligent wind power data and digital twin wind farm models, transcend the limitations of physical entities in the development of intelligent wind power technology. In this way, such technology is able to comprehensively drive far-reaching changes in wind power production, operation, maintenance, and management.

At present, the large-scale application of digital twin technology in the operation, maintenance, and management of wind farms is relatively limited. The business involved needs to continue to expand. It is still facing the problems of uneven levels of data collection capabilities in the enterprise and the industry, and the underlying key data cannot be effectively perceived. In addition, the low rate of data use and lack of integration of data correlation and data mining make it difficult to play the hidden value of data. In the long run, the full potential of digital twin technology will be realized by the effective integration of data from the bottom layer to the top level of a wind farm, and it is necessary to integrate all systems and data in the entire intelligent wind power ecosystem.

Chapter 8
Offshore Wind Power and Comprehensive Utilization of Offshore Areas

8.1 Introduction

Due to the scarcity of offshore area resources, in recent years, China has proposed the concept of intensive use of the sea. As a result, offshore wind power and comprehensive utilization of offshore areas should be considered seriously. The comprehensive utilization of offshore wind power and seawater desalination, hydrogen energy, marine ranching, and other energy sources can improve the utilization efficiency of offshore areas and project revenue. It is an important trend in the future development of offshore wind power [1–6].

Combining with offshore wind farms, a major offshore "energy island" demonstration project was built to integrate wind energy, hydrogen energy, seawater desalination, energy storage, marine ranching, etc. The project aims to provide coastal cities with high-quality, low-cost, and pollution-free electricity. At the same time, the use of energy storage and offshore wind power to achieve multi-energy complementarity will provide more revenue-increasing ways for offshore wind power to realize the cross-field application of wind power, and jointly promote green development and cost reduction.

8.2 Offshore Wind Power and Seawater Desalination

China is seriously short of freshwater resources and the distribution of freshwater resources is uneven in time and space. Considering population growth, increasing industrialization, and urbanization, China's demand for freshwater is more urgent. Seawater desalination is an important way to solve the shortage of freshwater resources for human beings. It is an important supplement and strategic reserve for freshwater resources.

© The Author(s), under exclusive license to Springer Nature Singapore Pte Ltd. 2025
Y. Liu et al., *Offshore Wind Farm Technology*,
https://doi.org/10.1007/978-981-96-1889-7_8

At present, China's seawater desalination projects are mainly distributed in coastal areas with severe water shortages, such as Tianjin, Zhoushan, Qingdao, Dalian, and other coastal cities. According to the goal of the "13th Five-Year Plan for National Seawater Utilization", by 2020, the total scale of seawater desalination in the country will reach 200,000 t/day, and the newly added seawater desalination scale in coastal areas will exceed 1.05 million t/day.

The desalination scale is more than 140,000 t/day. It is estimated that by 2023, the water production scale of seawater desalination will reach 2.85 million tons per day, and the development of the seawater desalination industry will reach a new level.

In recent years, with the global emphasis on seawater desalination technology, seawater desalination technology has shown a trend of rapid development. Sustainable clean energy such as solar energy, wind energy, and nuclear energy has been valued due to their large reserves, environmental friendliness, and sustainable utilization. The application of new energy for seawater desalination has become a global development trend. This is also an inevitable requirement to ensure China's energy and resource security and sustainable development.

8.2.1 Desalination Technology

Seawater desalination is the process of removing most of the salts in seawater to make the treated seawater meet the standards of residential and industrial water use. In short, it is the process of obtaining fresh water from seawater. At present, the mainstream technologies of global seawater desalination can be divided into two categories: the first category is heat-driven technology, which uses heat-based distillation to separate fresh water, mainly including low-temperature multi-effect distillation (MED) and multi-stage flash evaporation (MSF) and vapour compression distillation (VC), etc.. It has the advantages of system stability, reliability, and high water quality, but the disadvantage is high energy consumption; the second category is membrane technology, which consumes mainly electricity, including reverse osmosis technology (RO) and electrodialysis (ED) have the advantages of low one-time investment, low energy consumption, and high operating flexibility. The main disadvantage is that the maintenance cost is relatively large [1].

Both heat-driven technology and membrane technology are mature technologies that have been applied in large-scale engineering. In terms of application scenarios, seawater desalination plants that have been operating on a large scale in the Middle East mainly use MSF and MED technologies. Australia, located in Oceania, mostly uses RO technology in its seawater desalination plants [6].

The main feature of seawater desalination technology is high energy consumption. The energy consumption of conventional surface water treatment is about 0.06 kWh/t, while the average energy consumption of seawater desalination is about 8.4 kWh/t, which is 140 times more than that of conventional water treatment, about 40% of the total cost [5].

Table 8.1 Energy consumption of seawater desalination technology

	MSF	MED	TVC	MVC	RO	ED
Scale/(10,000 t/d)	5 ~ 7	0.5 ~ 1.5	1 ~ 3	0.01 ~ 0.25	9.8 ~ 12.8	2 ~ 14.5
Power consumption/ (kW h/t)	4 ~ 6	1.5 ~ 2.5	1.6 ~ 1.8	7 ~ 12	1.5 ~ 5.5	0.7 ~ 5.5
Thermal energy consumption/ (kW h/t)	15.83 ~ 21.5	12 ~ 19	14	0	0	0
Total energy consumption/ (kW h/t)	19.83 ~ 27.5	13.5 ~ 21.35	16 ~ 18	7 ~ 12	1.5 ~ 5.5	0.7 ~ 2.5 2.64 ~ 5.5

The energy consumption of seawater desalination technology is shown in Table 8.1: the average energy consumption of the desalination process dominated by heat-driven technology is about 20.5 kWh/t; the desalination process dominated by membrane technology has a relatively low average energy consumption of about 3 kWh/t when producing the same amount of fresh water. As the most energy-saving desalination technology, reverse osmosis accounts for more than 60% of the total desalinated water at present. The energy consumption of the device is 1.5–5 kWh/t, and the energy consumption accounts for 39–50% of the total cost of water production.

In summary, the energy consumption of seawater desalination directly determines the cost. Higher energy consumption increases the total cost of freshwater production and reduces the effect of energy conservation and emission reduction through seawater desalination.

8.2.2 *Synergistic Development of Offshore Wind Power and Seawater Desalination*

(1) **Advantages of synergistic development**

The development of renewable energy desalination is the best way to solve the dilemma of the shortage of freshwater resources, the depletion of fossil energy, and the deterioration of the ecological environment. Based on the distribution of wind resources and the shortage of freshwater resources in China, it can be seen that the synergistic development of offshore wind power and seawater desalination has a good match in the southeast coastal areas of China. Complementary resource advantages can be formed, and the abundant wind energy and seawater resources in coastal areas can be used according to local conditions. While increasing the utilization rate of renewable energy, it provides an effective way to alleviate the shortage of fresh water in adjacent areas.

Compared with fossil fuel energy, offshore wind power can reduce carbon emissions caused by seawater desalination, while also solving the problem of wind curtailment. Onshore wind power consumes a lot of land resources and has a certain impact on soil and water conservation and vegetation. Compared with onshore wind power, offshore wind power highlights geographical and environmental advantages, and is easy to couple with seawater desalination systems. For non-grid-connected offshore wind power, it can also provide energy support for seawater desalination systems while solving the electricity consumption of island residents [4].

(2) **Synergistic development pattern**

Currently, wind energy can directly or indirectly power desalination systems through four energy media: electrical energy, heat energy, gravitational potential energy, and kinetic energy.

- Electrical energy. This is the most commonly used form of energy in the coupling process of offshore wind power and seawater desalination systems. Offshore wind power converts wind energy into electrical energy to drive seawater desalination systems. For grid-connected offshore wind power, when the wind power output is insufficient, the minimum production demand for seawater desalination is given priority; when the wind power output is surplus. Surplus wind power can be consumed as much as possible. For off-grid offshore wind power, the intermittent nature of wind power must be considered. When the wind speed exceeds or cannot meet the requirements, backup facilities (such as batteries, pumped storage, and flywheel systems) need to be integrated into the system to store or release energy. This method is mainly aimed at islands where freshwater resources are scarce and are used to solve problems such as the inconvenience of conventional energy supply.
- Heat energy. The wind energy is directly converted into frictional heat energy to heat the distiller, and the structure is simple.
- Gravitational potential energy. To reduce the energy loss caused by wind energy conversion, gravitational potential energy is also used as a medium between wind energy and seawater desalination process, such as converting wind energy into gravitational potential energy stored in a certain height reservoir by wind turbines.
- Kinetic energy. On Coconut Island off the northern coast of Oahu, Hawaii, a brackish water desalination wind-powered reverse osmosis device has been established. The system is directly driven by the shaft power of the wind turbine through a high-pressure pump.

(3) **Cases of synergistic development**

At present, there are many demonstration projects integrating offshore wind power and seawater desalination systems in China: Offshore wind power + seawater desalination fusion test and demonstration application base in Qingdao deep sea and far sea 2 GW offshore wind power integration demonstration wind farm project;

Huizhou explores the application research of island independent power system integrating wind energy, solar energy, wave energy, and other power generation, promotes the industrial application of seawater desalination system and technology, sets up seawater desalination research and development bases in areas with abundant seawater resources and abundant power resources, and explores a new model of "hydropower cogeneration"; Based on the efficient utilization of resources, Yancheng established the application mode of "wind power and water", and built the first non-grid wind power seawater desalination production line in the country. Among them, the complete set of seawater desalination technology for the independent operation of intelligent micro-grid wind power is the first in the world. The input is seawater and the outflow is fresh water that meets the standard of bottled purified water (purified water for drinking) [2]. At the beginning of 2021, China's first nuclear power integrated smart energy project will start construction. This project is a new future model of the energy industry constructed by SPIC, which centrally collects and monitors multiple energy sources such as nuclear energy, solar energy, wind energy, and energy storage, and realizes multi-energy complementarity. Among them, the offshore wind power project will create a three-dimensional development demonstration project of marine resources integrating offshore wind power + seawater hydrogen production + sea ranching + sea sightseeing.

(4) **Coordinated development challenges**

Seawater desalination using wind energy is a cross-field (multi-disciplinary) application of offshore wind power, and there are still some problems: First, wind energy is intermittent and unstable, how to adapt traditional seawater desalination technology to wind energy is the key to the development of offshore wind power desalination projects. Secondly, the electricity price of offshore wind power is also the main factor restricting its cross-field application to seawater desalination. It is still difficult to achieve grid parity while meeting the profitability requirements of wind power developers [3].

8.3 Offshore Wind Power Product Hydrogen

8.3.1 Strengths and Challenges

Hydrogen energy is a new type of energy, which has the advantages of high fuel calorific value, abundant resources, and non-polluting combustion products. It has the reputation as "the ultimate energy in the twenty-first century". At present, global hydrogen energy is on the fast track (fast lane) of development. Many countries attach great importance to the development of hydrogen energy and have issued hydrogen energy development strategies or industry development guidelines.

According to the World Energy Council, hydrogen is divided into three categories: grey hydrogen, blue hydrogen, and green hydrogen [7]. Among them, grey hydrogen is produced hydrogen through fossil fuels (Grey hydrogen is hydrogen gas produced from fossil fuels); blue hydrogen is produced hydrogen (Blue hydrogen is hydrogen gas produced through) steam methane reforming technology or coal gasification plus carbon capture and storage technology; Green hydrogen is to use renewable energy to electrolyze water to produce hydrogen, which can be truly realized zero emission of carbon dioxide, be friendly to the environment, and have the highest social acceptance. The development of green hydrogen is also the first choice to realize the vision of carbon peak and carbon neutrality.

In February 2021, according to the latest report released by Bloomberg NEF, currently announced electrolytic hydrogen production projects are mainly divided into three types according to the type of power source: solar power, onshore wind power, and offshore wind power. Other types include Grid electricity, nuclear power, hydropower hydrogen production, etc.

With China announcing the carbon neutrality goals, how to solve the problem of large-scale new energy power grid integration and consumption has become an urgent need. With the development of hydrogen production technology and hydrogen storage technology, new energy hydrogen production technology represented by wind power and photovoltaic power generation hydrogen production technology has gradually matured and has the conditions for industrialization. However, China's wind power hydrogen production and solar hydrogen production are facing some challenges.

(1) **Solar hydrogen production**

For China's solar hydrogen production, there are two important limitations:

- The areas with rich solar energy in China usually lack water resources, and the water consumption for hydrogen production is insufficient. From the perspective of pure stoichiometry, it takes 9 L of water to prepare 1 kg of green hydrogen, which cannot achieve the ideal efficiency in the actual production process. Taking into account the water desalination process and additional water consumption, the average water consumption for producing one kilogram of green hydrogen will reach 18–24 kg.
- The water consumption of solar hydrogen production is higher than that of wind energy, and when the electrolyzer is coupled with photovoltaics, the water consumption of photovoltaic hydrogen production can reach 22–126 L (depending on solar radiation, lifetime, and silicon content). This further limits the feasibility and economics of solar hydrogen production. Long transport distance [8].

The western region with the most abundant solar energy resources is far away from the coastal economic centre, which means higher transportation costs for green hydrogen. Regardless of pipeline or rail transportation, it means that a large amount of investment in hydrogen transportation infrastructure is required.

(2) **Hydrogen production by onshore wind power**

China's onshore wind power hydrogen production has certain cost advantages. However, in the long run, the growth space for onshore wind power hydrogen production will be relatively limited. This is because the preparation of green hydrogen often requires 1.3–1.8 times the new energy power generation resources, that is, a 1 GW green hydrogen plant needs to be equipped with a 1.3–1.8 GW new energy power station. Such a resource demand requires a relatively large space for onshore wind resources, which will become an increasing challenge in the future when the development of onshore wind power projects is gradually slowing down.

(3) **Offshore wind power hydrogen production**

Offshore wind power hydrogen production provides an opportunity to realize the efficient utilization of marine resources in multiple ways. Offshore wind power hydrogen production has three major advantages:

1) Offshore wind resources are abundant, the capacity coefficient of wind power is higher than that on land, and the efficiency of hydrogen production is higher than that of solar energy and land wind power;
2) The space resources of offshore wind power are vast, especially when considering the future deep-sea floating wind power, there will be a very large room for development in the future;
3) Offshore wind power resources are close to China's main economic centres, and the transportation cost of hydrogen production from offshore wind power is relatively low.

The electricity cost is the main component of green hydrogen production cost. Due to the late start of offshore wind power in China, the cost is higher than that of solar energy and onshore wind power. It is generally believed that based on the current market hydrogen price of 70 yuan/kg, the electricity cost of hydrogen production by electrolysis of water needs to be controlled within 0.3 yuan/kWh to achieve economical hydrogen production. At present, the guide price of offshore wind power is as high as 0.75 yuan. Under the current power policy, hydrogen production from wind power is less economical than fossil fuel cell hydrogen production such as coal hydrogen production [9, 10].

On the whole, with the gradual decline in the cost of offshore wind power, the gradual realization of grid parity for offshore wind power, and the gradual expansion of the demand for green hydrogen energy. The hydrogen production from offshore wind power will have a broader market prospect.

8.3.2 *Offshore Wind Power—Hydrogen Energy Integrated Energy System*

Offshore wind power hydrogen production, that is, the electricity generated by offshore wind power is converted into hydrogen gas through water electrolysis hydrogen production equipment and transported to the place where hydrogen is used. The specific process is as follows: wind power generation-hydrogen production by electrolysis of water-hydrogen storage and transportation-applied to various industries, such as transportation, industrial thermal processing, chemical industry, etc. [11].

The Guangdong Electric Power Design and Research Institute proposed the technical route and system composition of offshore wind power hydrogen production including offshore hydrogen production stations and onshore hydrogen refuelling stations.

Offshore wind power-hydrogen energy comprehensive energy system includes seawater desalination device, water electrolysis hydrogen production device, compressed hydrogen storage device, wind turbine monitoring system and supporting electrical access devices, etc. The hydrogen production system is integrated into the offshore booster station. At the same time, the hydrogen storage and refuelling parts are arranged in the onshore centralized control centre. Figure 8.1 is a schematic diagram of an offshore wind power-hydrogen-integrated energy system.

Onshore hydrogen refuelling station: Including high-pressure hydrogen storage unit and hydrogen decompression distribution plate. The high-pressure hydrogen

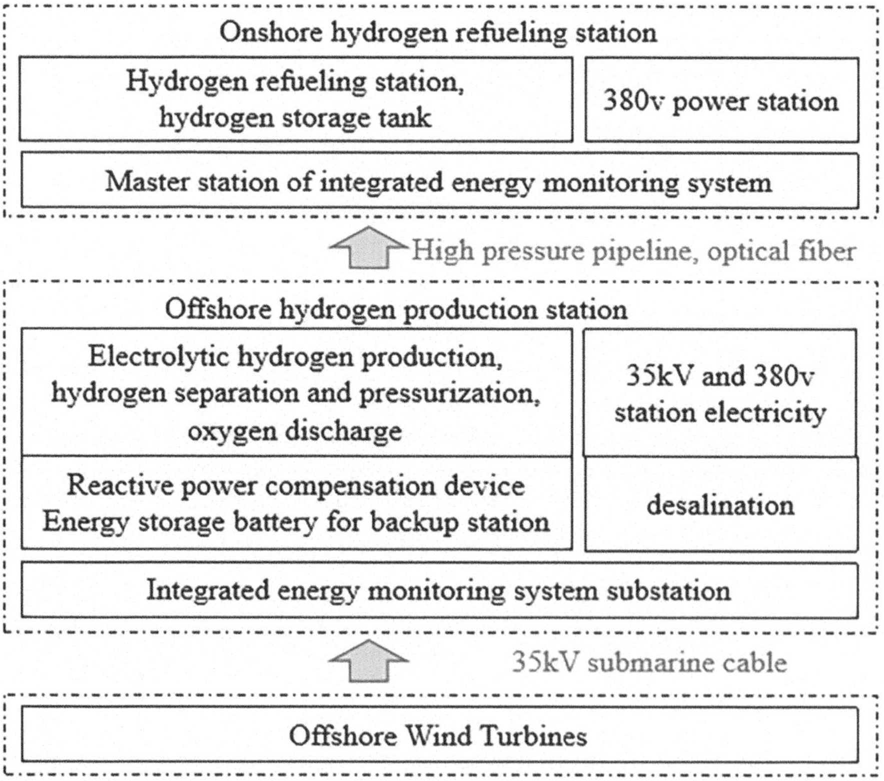

Fig. 8.1 Schematic diagram of offshore wind power-hydrogen energy integrated energy system

storage system is to store the compressed and pressurized hydrogen in the high-pressure hydrogen storage bottle group of the alkaline electrolyzer hydrogen production system, and the hydrogen storage tank is installed outdoors. The decompression distribution plate is for the user to obtain decompressed hydrogen from the hydrogen storage tank, and is equipped with a safety valve.

Offshore hydrogen production station: By receiving the electric energy generated by wind turbines, hydrogen is generated in the electrolyzer, and high-purity hydrogen is produced through steps such as separation, drying, and purification. High-purity hydrogen is pressurized and sent to onshore hydrogen refuelling stations through pipelines. The water electrolysis hydrogen production system includes a water electrolyzer, seawater desalination, hydrogen purification device, and hydrogen compressor, etc., the oxygen produced is directly discharged (and the oxygen generated by it is directly discharged).

Offshore wind turbines: They can accept commands from the onshore comprehensive energy monitoring system, and automatically adjust and control the energy output capacity of the wind turbines in the wind farm according to the pre-agreed control strategy, to finally realize the active and reactive power control of the wind farm.

Integrated energy monitoring system substation: It is necessary to ensure the safe operation of wind turbines and maximize the benefits of hydrogen production. It is mainly composed of an automatic power generation control subsystem and an automatic voltage control subsystem to realize the scheduling and control of the entire wind farm.

The hydrogen production link of this scheme is at sea. At present, most of the existing pilot projects are hydrogen production in onshore electrolyzers. Each design has pros and cons.

8.3.3 Discussion on Offshore Wind Power Hydrogen Production Model

With the rapid development of global offshore wind power, the lagging power grid construction speed cannot meet the rapidly expanding demand for offshore wind power transmission, which has become a dilemma for the rapid development of offshore wind power in European countries. China's offshore wind power is facing problems such as grid parity, rush installation, technological innovation, and overall planning. After large-scale offshore wind power was put into production,(After the production of large-scale offshore wind power), how to solve the grid connection and consumption of offshore wind power has become a problem that must be solved both at home and abroad. In this context, the use of offshore wind power to prepare hydrogen and send it to the hydrogen energy market through various storage and transportation technologies, and the development of cross-power transmission channels provide a feasible idea for the development of offshore wind power. For different application scenarios, offshore wind power hydrogen production models are different. Through the existing offshore wind power hydrogen production projects in the world in recent years and the projects that may be developed and constructed in the future, the mode of offshore wind power hydrogen production will be discussed from the perspectives of project design and operation mode.

From the perspective of project design, offshore wind power hydrogen production can be roughly classified into the following two modes [12].

(1) **Onshore hydrogen production**

At present, the most common project design is to transmit the power generated by offshore wind turbines to land through submarine cables and then produce hydrogen by onshore electrolyzers. The advantage is that the technical risk is relatively small, and the scale of hydrogen transportation is also small. The electrolyzer obtains electricity from different sources, which can be divided into three schemes:

- High-voltage grid connection, the offshore wind power is transmitted to the high-voltage grid through the grid, and then the grid transmits power to supply power for the electrolyzer.
- The offshore wind farm and the grid supply power to the electrolyzer at the same time. The electrolyzer can obtain power from the special cable of the offshore wind farm or through the grid. The electrolyzer is built on land. This scheme is not only conducive to reducing the impact on the transmission grid but also can reduce grid costs and electricity prices from two aspects of power transmission and distribution.
- The electrolyzer is directly coupled to the offshore wind farm. The electrolyzer is built by the offshore wind farm and is directly coupled with the wind farm, and the electricity from the offshore wind power is directly delivered to the onshore electrolyzer. The advantage of this solution is that the electrolyzer can directly obtain electricity at a cost price, but it needs policy support and cannot completely solve the problem of transmission grid construction.

(2) **Offshore hydrogen production**

At present, there are two common design schemes: deploying electrolyzers offshore and transporting hydrogen through pipelines, the details are as follows:

① The electrolyzer is directly coupled with the offshore wind farm, which is similar to the second scheme of onshore hydrogen production, the difference is that the electrolyzer is built on an offshore platform.

② Offshore hydrogen production by electrolysis. The electrolyzer is built inside or on the platform of the offshore wind turbine, replacing the offshore cable with a lower-cost transmission method, and obtaining electricity at the cost price of the wind farm. This method needs to solve the problem of transportation from sea to land after hydrogen production. There are already projects planning to use offshore oil and gas pipelines to transport hydrogen. As offshore wind farms move farther and farther from shore, the cost of transmitting electricity has also increased. With the help of the ready-made oil and gas pipeline network, the investment in outgoing cables is saved, and it becomes the most potent option to continue to significantly reduce the cost of offshore wind power in the future. There are currently two mainstream options:

- First, build a special offshore platform to collect the power of wind turbines, install electrolysis equipment on the platform to produce hydrogen, and then send it out through pipelines. (The first option is to build a special offshore platform to collect the power of wind turbines, install electrolysis equipment on the platform to produce hydrogen, and then send it out through pipelines.)
- Second, expand the conventional wind turbine platform, equip the wind turbine platform with hydrogen production equipment, send it directly through pipelines, and completely abandon the link of cable transmission, (The second option is to expand the conventional wind turbine platform, equip the wind turbine platform with hydrogen production equipment, send it directly through pipelines, and completely abandon the link of cable transmission,) which is the current key research direction for offshore wind power hydrogen production: On the one hand, it can effectively alleviate the impact of offshore wind power on the onshore power grid after large-scale production; on the other hand, it can realize energy transformation. It is necessary to point out that China already has a high level of power grid construction and rich experience in construction with a well-developed power grid. The demand for off-grid electrolysis is limited, and there is also a clearer (clear) demand for the loosening of power grid policies.

From the perspective of operation mode, offshore wind power hydrogen production can be roughly classified into the following two modes:

- Hydrogen production from surplus electricity. Wind power configuration with hydrogen energy storage can smooth wind power output and improve consumption capacity. Most planned projects envisage simultaneous electricity supply and hydrogen production. Since the green hydrogen market is still in its infancy, such projects do not have a mature business model and are still in the pilot stage.
- Mainly hydrogen production. Hydrogen is obtained by electrolyzing seawater at sea and then transported to land by using existing oil and gas pipelines with sufficient capacity. This not only does not require the laying of long-distance submarine cables but also can share pipeline maintenance costs with oil and gas developers in the future. Promoting hydrogen production from offshore wind power in sea areas rich in wind resources and oil and gas resources can promote energy transformation. Offshore wind power projects do not require new offshore transmission systems and are not subject to the constraints of grid companies. Wind farms are built near oil and gas platforms or oil and gas pipelines to reduce transmission loss and project investment costs. Seawater, the raw material for hydrogen production, is inexhaustible. Although this pattern has many advantages, its development needs further research, such as offshore hydrogen production and storage technology, environmental impact issues, and safe coexistence with gas platforms. In

addition, how to develop the business pattern needs to be further studied, such as offshore hydrogen production and storage technology, environmental impact issues, and safe coexistence with gas platforms. How to develop the business model also needs further study.

8.3.4 Global Offshore Wind Power Hydrogen Production Projects and Development Direction

With the rapid development of global offshore wind power, the pace of construction of offshore wind power hydrogen production projects at home and abroad is also following up. Major offshore wind power construction countries are actively exploring innovative models for offshore wind power hydrogen production [13, 14].

In October 2018, the National Development and Reform Commission and the Energy Administration issued the "Clean Energy Consumption Plan (2018–2020)", proposing to "explore the conversion of surplus electricity from renewable energy into heat, cold, and hydrogen energy, and realize the multi-channel approach to renewable energy. Efficient utilization" provides a feasible idea for the development of China's offshore wind power hydrogen production. At present, there are a few demonstration projects of offshore wind power hydrogen production in China, but they have not entered the stage of large-scale application.

In 2009, State Grid Shanghai Electric Power Co., Ltd. took the lead in launching the project of "preliminary research on hydrogen production technology combined with wind power and seawater", and proposed various application schemes for hydrogen production by wind power and photovoltaic power generation. Taking the East China Sea wind farm as an example, the benefit evaluation of wind and solar hydrogen production was carried out, which laid the foundation for the subsequent research on offshore wind power hydrogen production technology.

In June 2020, as the first national-level deep-sea and deep-sea integrated demonstration wind farm project, the Qingdao Deep-sea and Far-Sea 2 GW Offshore Wind Power Integration Demonstration Wind Farm Project made new progress. The direct investment in the wind power sector is more than 30 billion yuan, which can drive a total investment of more than 50 billion yuan in related industries such as the integration of wind farms and marine ranching industries, wind power hydrogen production, wind energy desalination, and equipment manufacturing. The project is expected to carry out experiments and demonstrations on the integration of offshore wind power + marine ranch, the foundation of floating wind turbines, long-distance offshore power transmission, hydrogen production from surplus electricity, and seawater desalination. The project aims to carry out experimental research on the first (set) of new wind power equipment, promote the demonstration and application of innovative floating offshore wind turbines in deep and far seas, comprehensively carry out the integration of offshore wind power + marine ranch applications and new technical equipment, etc. Besides, it aims to promote diversified integration experiments and demonstration applications of offshore wind power + wave power generation, offshore wind power + hydrogen production and hydrogen storage, offshore wind power + seawater desalination, offshore wind power + marine chemical industry, offshore wind power + marine scientific research, and create a world-class "offshore wind power +" A demonstration base for fusion projects.

In October 2020, the Guohua Investment Jiangsu Dongtai 0.5 GW Offshore Wind Power Project, a joint venture between China Energy Group and EDF, was unveiled and the ceremony for deepening Sino-French cooperation was held in Nanjing, marking China's first Sino-foreign joint venture offshore wind power project The official landing also shows that China's offshore wind power market has officially opened the door to foreign investment.

In October 2020, the unveiling ceremony of Guohua Investment Jiangsu Dongtai 0.5 GW Offshore Wind Power Project, a joint venture between China Energy Group and EDF, was held in Nanjing. It marks the official landing of China's first Sino-foreign joint venture offshore wind power project, and also indicates that China's offshore wind power market has officially opened the door to foreign investment.

In November 2020, Tongji University won the bid for CNOOC's offshore hydrogen production process technology research project, which aims to study, design and optimize the process flow of offshore wind power hydrogen production, and propose boundary conditions for technical and economic feasibility. The main contents of the project are: offshore electrolysis water hydrogen production process scheme selection and technical research; offshore wind power and hydrogen production equipment matching research; offshore hydrogen storage and hydrogen transmission technology research.

With the further development of domestic offshore wind power and the accumulation of experience in wind power hydrogen production projects, offshore wind power hydrogen production projects will also usher in (be on the threshold of) a development climax.

Europe is also vigorously developing offshore wind power hydrogen production. At the beginning of 2020, the European Union issued a hydrogen energy strategy, making it clear that vigorously developing green hydrogen is a core and important measure to achieve carbon neutrality. The North Sea of Europe has the best offshore wind resources in the world. As the development of offshore wind power in Europe gets farther and farther (further and further) from the coast, the water depth is getting deeper and the scale is getting bigger and bigger. After more than 60 years of offshore oil and gas development, the North Sea subsea natural gas pipeline network is densely covered, which can make full use of the subsea natural gas pipeline network to transport hydrogen. Therefore, offshore wind power will be the most important source of green hydrogen in Europe in the future. At present, the mainstream of the foreign offshore wind power hydrogen production market is still a pilot project, which has not entered the stage of large-scale commercial application.

As a leader in offshore wind power, the UK plans to build the world's largest offshore wind power hydrogen production project. Ørsted's 1.4 GW Hornsea II offshore wind farm will be connected to the Gigastack project to produce green hydrogen. The wind farm is scheduled to be completed and be put into operation in 2022 and it will replace the 1.2 GW Hornsea I as the world's largest offshore wind farm. Another green hydrogen project, Dolphin, plans to develop a 4 GW floating wind farm in the North Sea, using a 10 MW model, installing a hydrogen production sub-unit on each wind turbine, and sending hydrogen out through pipelines. According to the plan, the final investment decision of the prototype project will be completed by the end of 2021, and it will be put into operation in 2023, and the hydrogen production on the 10 MW model will be realized by 2026.

Germany has the world's second-largest installed capacity of offshore wind power, after the United Kingdom, which ranks first. Behind the rapid development of offshore wind power is the contradiction between the rapid growth of offshore wind power and the slow pace of grid construction. Germany will phase out nuclear power, and coal-fired power generation by 2038, and be committed to the development of renewable energy. In Germany's "National Hydrogen Energy Strategy", hydrogen production from offshore wind power has become one of the key development directions of green hydrogen. In 2019, refiner Raffinerie Heide launched the Westküste 100 project in Schleswig–Holstein in northern Germany with partners such as Thyssenkrupp, EDF (Germany), Open Grid Europe (OGE), OGE, and Ørsted. The project's green hydrogen will be produced from excess electricity generated by nearby offshore wind farms, with the goal of planning and realizing regional hydrogen commercialization on an industrial scale. Two companies under Engie, Tractebel Engineering and Tractebel Overdick, will build a 400 MW offshore wind power hydrogen production station in Germany. Recently, Siemens Gamesa and Siemens Energy announced cooperation to integrate their respective developments in the field of hydrogen production from wind power into an innovative solution. This solution seamlessly integrates the electrolyzer system into the wind turbine, and finally realizes the integration of the electrolyzer array at the bottom of the offshore wind turbine tower to produce hydrogen, and directly produces green hydrogen with a stand-alone coordinated system, opening up new ideas for offshore green hydrogen production. The two parties plan to fully demonstrate the offshore wind power hydrogen production system in 2025 or 2026.

In order to achieve the national environmental protection targets in 2030, the Netherlands plans to phase out coal-fired power and stop domestic natural gas production by 2022. The long-term plan of the Netherlands government is to mainly use green hydrogen and use offshore wind power to produce hydrogen as one of the key directions for future energy supply. The PosHYdon project based on the Q13a platform is the world's first offshore wind power hydrogen production project. It is jointly undertaken by a number of companies and institutions in the Netherlands to promote emission reduction and establish a new energy model in the North Sea. The project aims to research innovative technologies for hydrogen production from offshore wind power and promote their application in the Netherlands. Neptune Energy's Q13a platform is the first fully electrified oil and gas platform in the Dutch North Sea (North Sea of Netherlands) and will be transformed into a hydrogen production platform in the PosHYdon project. The container-type hydrogen production equipment is small in size and can be accommodated by most offshore platforms. By combining the three industries of hydrogen, natural gas and wind power, energy transformation is realized. On February 27, 2020, Dutch Shell announced the launch of Europe's largest offshore wind power hydrogen production project (NortH2). The NortH2 project plans to build a large-scale hydrogen production plant in Eemshaven, the Netherlands, to convert offshore wind power into green hydrogen. At the same time, it will establish an intelligent transportation network in the Netherlands and North-west Europe, and use green hydrogen for industrial and consumer markets through Gasunie's natural gas infrastructure.

In the first half of 2020, French energy giant Total (As the energy giant in French), Total launched a research project called the O/G Decarb Innovation Project. In addition to Total, participating companies or institutions include Danish marine energy company Floating Power Plant A/S, Danish Hydrocarbon Research and Technology Center, the world's largest wind energy public research centre Denmark DTU Wind Energy, hydrogen energy research institution Hydrogen Valley, Denmark The natural gas technology centre, etc., aims to explore the comprehensive utilization of floating offshore wind power, wave energy, hydrogen energy and other forms of energy as a power supply model for offshore oil and gas platforms.

In 2020, Belgium released a plan for an offshore wind power hydrogen production project called Hyport Oostende, which was jointly developed by the offshore engineering giant DEME, the investment institution PMV and the Belgian Ostend Port, and implemented in the Ostend Port. The project is divided into two phases: in the first phase, a 50 MW demonstration project will be developed; in the second phase, a larger-scale commercial project will be developed and completed by 2025. It is expected to become the world's first commercial offshore wind power hydrogen production project put into operation.

Offshore wind power projects in cooperation between Asia and Europe have also emerged. In 2020, Singaporean energy company Enterprize Energy and Engie's engineering company Tractebel Overdick signed a memorandum of understanding to jointly develop and build a "one-stop" offshore platform in the North Sea, using offshore wind power to produce hydrogen and ammonia. The platform is called Energy-Plus, with a planned capacity of 400 MW. The bottom is the jacket foundation, and the upper platform is equipped with electrolyzer equipment to produce green hydrogen and send it out through pipelines. At the same time, it also produces a small amount of green ammonia and uses gas tanks. Stored and transported uniformly.

On the whole, it is forward-looking, necessary, inevitable and dominant to vigorously develop offshore wind power hydrogen production in Europe. Excellent offshore wind resources, existing oil and gas pipeline resources extending in all directions and the global leading position of offshore wind power have laid the foundation for the development of offshore hydrogen production in Europe. In the future, driven by the low-carbon hydrogen production plans of many European countries, green hydrogen will increase rapidly, which will inevitably drive the growth and investment of the entire hydrogen supply chain. Europe has the opportunity to become a hydrogen energy technology innovation centre, hydrogen energy equipment manufacturing centre and hydrogen energy industry. Project construction, operation and maintenance centre.

All in all, the offshore wind power hydrogen production industry at home and abroad is in its infancy, and countries are still exploring feasible technical solutions and commercialization solutions. As a way of comprehensive utilization of marine energy, offshore wind power hydrogen production is expected to have a deeper and wider application space with the rapid development of offshore wind power, providing new ideas for offshore wind power consumption, offshore wind power transmission system construction, and energy transformation. As a country with large offshore wind power, hydrogen energy applications, and marine resources China should focus on the high-quality development of the offshore wind power industry, support and encourage more ways to tap the potential of marine resources and integrate with the hydrogen energy industry to solve the problem of green hydrogen sources. Promote the adjustment of China's energy structure to achieve the goal of carbon neutrality.

References

1. Guo J T, Wu S, Cui J et al (2017) Conceptual Design and Hydrodynamic Performance Analysis of an Integrated Wind Power and Seawater Desalination Platform. China Water Transport 17(12):140–142.
2. Liu C F, Li M, Wang Y Q et al (2019) Progress and application of seawater desalination technology. City and Town Water Supply 02:54–58+62.
3. Chen C, Yang Y, Wang Z et al (2019) Development Status and Prospect on Co-Development of Desalination and Clean Energy. Journal of Green Science and Technology 08:147–154.
4. Ahmed F E, Hashaikeh R, Diabat A et al (2019) Mathematical and optimization modelling in desalination: State-of-the-art and future direction. Desalination 469: 114092.
5. Ma Q, Lu H (2011) Wind energy technologies integrated with desalination systems: Review and state-of-the-art. Desalination 277(1–3):274–280.
6. Eke J, Yusuf A, Giwa A et al (2020) The global status of desalination: An assessment of current desalination technologies, plants and capacity. Desalination 495: 114633.
7. Luo Z X, Cao Y (2020). Development Prospect of Hydrogen Energy Industry and Its Development Path in China. Sino-Global Energy 25(02):9–15.
8. Franco B A, Baptista P, Neto R C et al (2021) Assessment of offloading pathways for wind-powered offshore hydrogen production: Energy and economic analysis. Applied Energy 286:116553.
9. Babarit A, Gilloteaux J C, Clodic G et al (2018) Techno-economic feasibility of fleets of far offshore hydrogen-producing wind energy converters. International Journal of Hydrogen Energy 43(15):7266–7289.
10. Yang X, Xu F (2016). Economical Model Analysis of Power to Gas. Journal of Low Carbon 5(04):37–42.
11. Yang Y, Chen L, Wang X H et al (2020) Design of Integrated Offshore Wind Power-hydrogen Energy Monitoring System. Southern Energy Construction 7(02):35–40.
12. Li Xue L, Yuan L (2022) Development Status and Suggestions of Hydrogen Production Technology by Offshore Wind Power. Power Generation Technology 43(02):198–206.
13. Tian T, Li Y X, Huang L et al (2021) Comparative Analysis on the Economy of Hydrogen Production Technology for Offshore Wind Power Consumption. Electric Power Construction 42(12):136–144.
14. Yan C, Huang S, Qu Y P (2022) Review on hydrogen production technology from offshore wind power to achieve carbon neutrality. Integrated Intelligent Energy 44(05):30–40.

The manufacturer's authorised representative in the EU is Springer Nature Customer Service Centre GmbH, Europaplatz 3, 69115 Heidelberg, Germany. If you have any concerns regarding our products, please contact ProductSafety@springernature.com

Printed and bound by CPI Group (UK) Ltd, Croydon, CR0 4YY

01/10/2026

02225921-0002